Globalizing Oil

Oil is the world's most important commodity. It is also its most politicized. Yet in Globalizing Oil Llewelyn Hughes shows that since the 1970s governments across the advanced industrial states liberalized oil markets by freeing prices, lowering trade barriers, and privatizing national oil companies. How did this come about? And why do some governments continue to intervene in support of domestic firms? In answering these questions, Hughes shows that the transformation of the international oil market in the 1970s changed the risks and opportunities facing firms. He also shows their ability to benefit was conditioned by previous attempts to shape the competitive landscape. Changes in oil market governance in the 1980s and 1990s can thus be understood as a temporal process, in which institutional legacies affected firm strategies towards liberalization. Hughes' study has important implications not only for the politics of oil, but also for the study of economic liberalization.

LLEWELYN HUGHES is Assistant Professor of Political Science and International Affairs at George Washington University.

Business and Public Policy

Series Editor:

This series aims to play a pioneering role in shaping the emerging field of business and public policy. *Business and Public Policy* focuses on two central questions. First, how does public policy influence business strategy, operations, organization, and governance, and with what consequences for both business and society? Second, how do businesses themselves influence policy institutions, policy processes, and other policy actors, and with what outcomes?

Other books in the series:

TIMOTHY WERNER, *Public Forces and Private Politics in American Big Business*

HEVINA S. DASHWOOD, *The Rise of Global Corporate Social Responsibility: Mining and the Spread of Global Norms*

Globalizing Oil

Firms and Oil Market Governance in France, Japan, and the United States

LLEWELYN HUGHES

CAMBRIDGE
UNIVERSITY PRESS

CAMBRIDGE
UNIVERSITY PRESS

University Printing House, Cambridge CB2 8BS, United Kingdom

Published in the United States of America by Cambridge University Press, New York

Cambridge University Press is part of the University of Cambridge.

It furthers the University's mission by disseminating knowledge in the pursuit of education, learning and research at the highest international levels of excellence.

www.cambridge.org
Information on this title: www.cambridge.org/9781107041998

First published 2014

Printed in the United Kingdom by CPI Group Ltd, Croydon CR0 4YY

A catalogue record for this publication is available from the British Library

Library of Congress Cataloguing in Publication data
Hughes, Llewelyn, 1971–
Globalizing oil : firms and oil market governance in France, Japan, and the United States / Llewelyn Hughes.
pages cm. – (Business and public policy)
Includes bibliographical references and index.
ISBN 978-1-107-04199-8 (hardback)
1. Petroleum industry and trade–Government policy–France. 2. Petroleum industry and trade–Government policy–Japan. 3. Petroleum industry and trade–Government policy–United States. I. Title.
HD9560.6.H84 2014
338.2'7282–dc23
2013028332

ISBN 978-1-107-04199-8 Hardback

Contents

List of figures and tables *page* viii
Preface ix

1 The puzzle of oil 1
The puzzle: explaining changes in oil market governance 5
An answer: firms and oil market governance 9
Firms and oil market governance 11
The role of government 15
The importance of oil market governance to the world economy 17
Overview of the book 19

2 Oil markets and the physiocratic fallacy 23
Changes in the international oil market 24
Emergence of a global oil market 27
Oil market governance in the advanced industrialized states 30
Defining change in oil market governance 32
Measuring change in oil market governance 33
Changes in oil market governance in the major oil-importing states 1980–2005 36

3 Explaining changes in oil market governance 39
Explaining changes in oil market governance 40
Transaction costs, power, and oil markets 41
Domestic interests and oil market governance 44
Firms, industrial compacts, and oil market governance 49
Firm interests and strategies 50
The origins of firm characteristics 55
Alternative explanations: the role of ideas and the influence of Europe 58
Empirical strategy 62
Challenges 64
Measuring oil market governance, and firm and policymaker strategies 65

4	Transforming French oil market governance	68
	Oil market governance in France	70
	The role of Europe	78
	Transforming oil market governance in France	81
	Markets and firm strategy	83
	Firm strategy and oil market governance	88
	Shifting policymaker strategies	95
	Summary	103
	Alternative explanation – the influence of Europe	104
	Alternative explanation – ideational change	110
5	Adjusting oil market governance in Japan	113
	Oil market governance in Japan	115
	Postwar oil market governance	122
	Adjusting oil market governance in Japan	127
	Markets and firm strategy	128
	Stable and unified firm strategies	132
	Shifting policymaker strategies	136
	Summary	144
	Alternative explanation – ideational change	146
6	The United States and energy independence	149
	Oil market governance to 1980: between conservation and production	154
	Promoting production	158
	Loss of competitiveness and rise of protectionism	161
	Oil market governance and firm characteristics	166
	Explaining changes in oil market governance since 1980	169
	Case study 1 – falling prices and oil market governance	171
	Stable but divided firm strategies	174
	Stable but divided policymaker strategies	181
	Case summary	184
	Case study 2 – rising prices and the George W. Bush administration	185
	Stable and unified firm strategies	188
	Stable but divided policymaker strategies	190
	Case summary	193
	Summary of changes in oil market governance in the United States	194
	Alternative explanation – ideational change	196
7	Firms, governments, and oil market governance	198
	Industrial compacts as a source of firm characteristics	199

Broader lessons for the study of economic liberalization 204
The politics of resource nationalism 207

Appendix: Oil market liberalization in the advanced industrialized states 210

Select bibliography 220

Index 247

Figures and tables

Figures

2.1 Crude oil production of the United States and the rest of the world (1918–44) *page* 25
2.2 Mean change in oil market governance (1980–2005) 38
2.3 Changes in oil market governance in France, Japan, and the United States (1980–2005) 38

Table

2.1 Measuring oil market governance 37

Preface

I became interested in the political economy of oil while working in Tokyo in the early 2000s. An important part of my job was persuading policymakers of the merits of liberalizing Japan's electricity and natural gas markets, and I found that security of energy supplies was often raised by regulators and policymakers as an important reason why liberalization would be unwise. Tracing the origins of Japan's energy insecurities led me to oil and what makes it different.

As I began work on the project I found that rather than being different, the choices governments have made over the past few decades about how to govern oil markets have much in common with the changes implemented in other sectors. Just as quotas and tariffs have been cut in manufactured goods, services, and other areas, and firms across a wide range of industries have been privatized, so the barriers to market entry in the oil industry have also fallen across the advanced industrialized states. Why this happened despite the historical treatment of oil as qualitatively different from other products, and why some countries continue to treat oil as a strategic resource, was the puzzle that motivated me to begin work on this study.

I was also surprised that there has been little written that documents and explains how the governance of oil markets across the OECD countries has changed since the wave of research that followed oil market volatility in the 1970s. This strikes me as an important gap in our knowledge, given the importance and enormous value of oil and oil products traded globally. Although there is surely more work to be done, this book is thus also partly an attempt to reincorporate energy politics into the study of comparative and international political economy in the advanced industrialized states.

The study was completed with the support of many people. Richard Samuels, Suzanne Berger, and Kenneth Oye at the Massachusetts Institute of Technology (MIT) deserve special thanks for their intellectual support. They have each influenced the approach I took towards the study of political economy and I am grateful for their guidance. At MIT, J. Patrick Boyd, Edward Cunningham, Benjamin Friedman, Michael Glosny, Brendan Green, Yinan He, Shirley Hung, Colin Jackson, Jon Lindsay, Austin Long, Mike Lynch, Gautam Mukunda, William Norris, Joshua Rovner, and Paul Staniland all helped me shape the ideas in the book. Laurie Scheffler, Casey Johnson, and Anita Kafka also helped during the development of the project. Most of all, I owe Nathan Cisneros a great deal for his support throughout the initial research and writing of the first draft.

The book manuscript took shape while at the Mossavar-Rahmani Center for Business and Government at the Kennedy School of Government. William Hogan has my thanks for hosting me as I began revisions, as does Louisa Lund for helping me navigate matters large and small at Harvard. I also thank Meghan O'Sullivan for introducing me to the work on energy policy being done at the Kennedy School.

I am lucky to have been able to complete the book at the George Washington University. I could not have asked for a better environment for writing. All my departmental colleagues have my thanks for their comradeship, but especially Charlie Glaser, Stephen Kaplan, Gina Lambright, Eric Lawrence, Kimberly Morgan, Harris Mylonas, Elizabeth Saunders, and Caitlin Talmadge. Susan Sell also deserves special thanks for organizing and hosting a workshop at the Elliott School of International Affairs, where I received valuable suggestions from Charlie Ebinger, Harvey Feigenbaum, Jonah Levy, Mike Mochizuki, Adam Sheingate, Mireya Solis, Emmanuel Teitelbaum, Kenneth Vincent, and others, each of whom read the manuscript and offered advice. Alain Beltran of the Centre National de la Recherche Scientifique read parts of the manuscript and offered detailed and helpful comments. Stephen Kaplan, Francisco Flores-Macias, David Victor, and Cornelia Woll also read and offered comments on versions of the argument. Henry Farrell deserves special thanks for his intellectual support and mentorship during my time at GWU. I would also like to thank Forrest Maltzman, Barbara Miller, Paul Wahlbeck, and most of all Stephanie Camillo for helping keep me sane as I managed some important bureaucratic issues. Excellent research assistance was provided by Jennifer Brea, Laura Franzen, Lara Lupu, Chris Mitchell, and especially Elsa Savourey, who went beyond the call of duty in helping gather materials. They each have my gratitude.

One of the best parts of research is learning how willing people are to lend their support, and how generous friends and strangers are with their time. In Tokyo I am particularly grateful to Kiyo Aburaki, Takeo Kikkawa, Ken Koyama, Masahisa Naito, Shigeru Sudo, Katsuhiko Suetsugu, and Toichi Tsutomu. Before and during my work in Paris I benefitted from the goodwill of Olivier Appert, Gilles Bellec, Dominique Charbit, Jean Marie Chevalier, Jean-Pierre Cordier, Nathalie Fokart, Erika Forzy, and Pierre Noel. In Washington, DC, Raelyn Campbell, Sarah Ladislaw, and Maureen Lorenzetti were generous with their help. I also thank all those interview subjects willing to share their experiences with me.

Institutional and financial support during the research for the project was provided by the Center for International Studies and the Laboratory for Energy and the Environment at MIT, the Smith Richardson Foundation, the Mossavar-Rahmani Center for Business and Government at the Kennedy School of Government, Harvard University, the Institute for Energy Economics Japan, the Japan Business Federation (Keidanren), and the Centre of Geopolitics of Energy and Raw Materials (CGEMP) at the Université Paris Dauphine. The final touches were made to the manuscript while I was a visiting fellow at the East-West Center, and I thank Satu Limaye for his hospitality.

I would like to thank John Haslam at Cambridge University Press. Aseem Prakash of the University of Washington and the series editor encouraged me throughout the submission and review process. I would also like to thank the two reviewers for their extended and excellent comments. I learned much from them, and their anonymous labors demonstrate the generosity that can make it so rewarding to be a part of this profession.

Lastly, I would like to thank Daniela and Leo, who put up with me throughout the research and writing of this book, and Tessa, who arrived part of the way through it. I dedicate this book to them, and to my parents, to whom I owe everything.

1 *The puzzle of oil*

Oil presents governments and firms with a challenge: how to govern the market for a commodity that is crucial to growth, security, and profits. Historically, governments identified scarcity – driven by dwindling reserves or the exercise of market power – as an important risk, and the rhetoric of government intervention continues to be cloaked in the language of national security. Yet the framing of oil as a strategic resource masks an important fact: over the past three decades, governments across the advanced industrialized states have abolished barriers to trade in oil and relinquished ownership in national oil companies that were often central to national strategies to manage risk in the international oil market.

This represents an astonishing turnaround. In some countries the laws and policies used to govern oil markets were established in the interwar years. Yet over the past three decades, governments have chosen to liberalize trade in crude oil and oil products, free prices, cut subsidies, and privatize firms in order to enable them to compete internationally on an equal basis.

These changes had important distributive implications. A simple calculation shows that at one hundred dollars per barrel, global spending on oil exceeds three trillion dollars per year. The International Energy Agency (IEA) estimates that one in every twenty dollars spent in global trade total will be spent on the buying and selling of crude oil in the next two decades. Decisions made in the major economies about how to govern oil markets thus redistribute billions of dollars between firms that produce and consume oil and oil products, and affect where oil is produced globally.

The changes in oil market governance documented in this book are also puzzling. For much of its history, trade in oil was controlled by a small number of firms based in the United States and Europe. The most important long-term implication of the volatility of the 1970s was the loss by these firms of the resources they had used to dominate markets globally. Today it is national oil companies based in the Middle East and elsewhere that own the vast bulk of the world's oil. Yet the response of governments across the advanced industrialized states was to abandon attempts to actively shape oil markets, even as some remain wedded to energy nationalism as one component of their energy security strategies.

These changes in oil market governance are also puzzling because a number of important explanations we have for why governments lowered the barriers to trade and investment in other areas of the economy are not readily applicable in oil markets. The international trade regime, for example, has played an important role facilitating the abolition of restraints on trade and investment by making it easier for governments to negotiate with one another, and increasing certainty that negotiating partners will honor agreements. Yet it has had limited influence in the markets for natural resources, including oil. Falling transport and communication costs are also identified as important facilitators of trade and investment. While there have certainly been extraordinary innovations in the oil sector, these have tended to enable firms to stay at home, rather than increasing their opportunities for exit. Indeed, many of these innovations were spurred by the nationalization of oil-producing assets in the 1970s, and focused on improving imaging of the subsurface, increasing oil recovery rates, and expanding offshore drilling opportunities with the goal of enabling firms to increase production in the Gulf of Mexico, the North Sea, and elsewhere.

It is also difficult to point to the role of economic power to explain falling barriers to trade and investment in oil markets. The United States, for example, is argued to have used a mix of positive and negative sanctions to prod open the national markets of its trading partners, with perhaps the clearest example being the pressure the US government applied to Japan in the 1980s and 1990s to deregulate its economy in order to reduce the bilateral trade deficit. Yet in the case of oil, efforts by the US government to open energy markets have been sporadic. Indeed, the United States has tended towards protectionism in oil markets, promoting the benefits of autarky under the rubric of energy independence, and subsidizing oil companies in order to increase domestic production as a share of total consumption.

These are by no means the only explanations for economic liberalization. We also need to take into account the influence of Europe, for example, as well as changing ideas about the appropriate role of governments in the market. The response of policymakers to the challenge of inflation and slowing growth in Europe and the United States is also important, and I consider each of these in the chapters that follow. Yet the relative unimportance of falling transaction costs and economic power raises the question of how we can explain changes made in oil market governance.

In this book I argue that an important reason for why governments adopted more liberal forms of oil market governance, and why some continued to protect domestic interests in the name of energy security, lies in the strategies of firms, and in particular how differences in the characteristics of firms conditioned how they responded to the emergence of a global market for oil.

The importance of firm characteristics in shaping national choices about how to govern oil leaves us with the question of where differences between firms come from. One of the puzzles in the international oil market is that most advanced industrialized states have negligible supplies of oil, putting them in a weak position within the oil economy. Yet despite this there are substantial differences in firm competitiveness cross-nationally. Even in the United States we see highly competitive firms operating alongside small independent oil producers that are weak in terms of their international competitiveness. And, as I show in this book, these differences have real implications for the positions they take in debates regarding oil market governance.

Where, then, do differences in the characteristics of firms come from? While managerial acuity surely accounts for some of this variation, I argue that in order to explain patterns of oil market liberalization today we need to understand how industrial compacts negotiated between business and governments in earlier periods shaped the characteristics of firms. Historical sequence therefore mattered, "priming" some firms to better manage the risks, and take advantage of the opportunities, that emerged along with the transformation of the international oil market, while limiting the ability of others to do so.

In the case of France, for example, which I take up in Chapter 4, I describe how the industrial compact negotiated between the international oil majors, local business interests, and the French government shaped the characteristics of firms operating in the oil sector. I then show that French oil firms responded to changes in the international oil market by seeking to further diversify operations internationally. Similarly, they began to reject rather than embrace government support as an instrument for securing market share and increasing profits.

In Japan, on the other hand, I describe how governments and firms sought to limit the dominance of the international majors. Yet in contrast to France, the industrial compacts negotiated between these groups allocated market share with little regard for whether domestic firms were likely to become industry leaders. Horizontal and vertical fragmentation, and ongoing protection, was the result. While liberalization of the domestic market was implemented, for reasons I document in Chapter 5, firms continued to focus on the benefits government could provide them as they sought to increase scale and profitability.

In Chapter 6 I extend the argument to the case of the United States. At first glance the politics of oil in the United States appears qualitatively different from the cases of France and Japan. Oil from the United States represented the bulk of oil produced globally until at least World War II, and the most powerful of the international oil firms were American in origin. Today the United States remains a crucial supplier to world markets, and this role is increasing thanks to the rise of unconventional oil. The case of the United States is nevertheless useful in two ways. First, while initial conditions in France and Japan were similar, there are important differences in political institutions in the two countries that could plausibly account for differences in outcomes between them. The case of the United States, on the other hand, allows us to consider how differences in the characteristics of oil firms shape their positions towards oil market governance, unencumbered by differences in political institutions. It therefore allows us another check as to whether the strategies adopted by firms in France and Japan stemmed from their underlying characteristics.

The case of the United States is also useful because it offers an opportunity to examine how resource endowments and these industrial compacts combined to shape the characteristics of firms. In Chapter 6 I describe how the importance of US oil in the global market helped large, integrated oil majors attain a dominant position in the world oil market prior to World War II. Yet, I also show that the US government continues to promote energy independence, channeling billions of dollars to domestic firms. This reflects, I argue, the interests of thousands of small- and medium-sized producers that also exist as a result of the historically privileged position of the United States in the international oil market.

My argument is not intended to supplant other explanations for the changes in economic governance in the major economies in recent decades. In the empirical chapters I record, for example, how a shift in the strategies policymakers used to govern markets in response to poor economic performance and government indebtedness also affected choices about oil market governance. But what I also show is that a historically informed account of the development of firm capabilities is crucial if we are to understand changes in market governance over the last three decades. Indeed, it is impossible to understand changes in oil market governance without taking into account firm strategies, and how these were affected by earlier attempts of business and governments to shape oil markets in their favor.

In the rest of this chapter I introduce these ideas in more detail. In the next section I outline some important explanations for changes in forms of market governance over the past three decades, and consider how well they help us to explain the two questions that motivate this study: how can we explain the liberalization of oil markets in the advanced industrialized states, and how can we explain variation in the extent to which governments in these countries liberalized? I then preface the main argument of the book, focusing on differences in the strategies firms adopted in response to changes in the international oil market, and the role of industrial compacts negotiated in earlier periods in shaping firm characteristics.

The puzzle: explaining changes in oil market governance

A central problem in political economy lies in explaining why countries transformed the rules used to govern economic markets over the past three decades. Part of the challenge lies in reconciling this with the conventional wisdom. Studies hold that reducing the role of government in actively shaping market outcomes improves welfare but has diffuse benefits. The benefits of protection, in contrast, are argued to be concentrated within particular industries or social classes. Logic therefore dictates that protection should be overprovided relative to what is considered socially optimal.[1]

[1] Vilfredo Pareto, *Manual of Political Economy* (New York: A.M. Kelley, 1971 [1927]), 379. For an attempt to reconcile theories of regulatory capture with liberalization see Sam Peltzman, Michael E. Levine, and Roger G. Noll, "The Economic Theory of Regulation after a Decade of Deregulation," *Brookings Papers on Economic Activity* (1989): 1–59.

If this is right, then how can we explain the wave of liberalization since the 1980s? One prominent answer points to falling transaction costs. The General Agreement on Trade and Tariffs, and World Trade Organization (GATT/WTO), for example, is understood to have reshaped trade policies by making it easier for governments to reach agreements. Even in the agricultural sector, where countries typically remain more protectionist, the linking of agricultural goods with other trade issues has enabled the gradual lowering of barriers to trade.[2]

What about the case of oil? While there is no explicit exemption for crude oil in the trade regime, governments have failed to reach agreement on how to treat natural resources within it. In part this is because important exporters were not members of the original General Agreement on Trade and Tariffs, and the United States became a net importer in 1948. This meant there was little pressure from exporters to force open markets in importing states. What limited attempts we have seen to incorporate natural resources in international trade rules have largely failed, most notably the New International Economic Order (NIEO) movement of the 1970s, and the efforts of the negotiating group on natural resource-based products during the Uruguay trade round.[3] In the case of oil, Article XXI of GATT/WTO also allows exclusions for reasons of national security, which provides a rationale for maintaining discriminatory trade policies given the historical framing of oil as a strategic resource rather than a simple commodity.[4]

2 On international organizations see Robert O. Keohane, *After Hegemony: Cooperation and Discord in the World Political Economy* (Princeton University Press, 1984), 49–64; Judith Goldstein, "International Institutions and Domestic Politics: GATT, WTO and the Liberalization of International Trade," in *The WTO as an International Organization*, ed. Anne O. Kreuger (University of Chicago Press, 1998), 133–52. On agriculture see Christina Davis, *Food Fights over Free Trade: How International Institutions Promote Agricultural Trade Liberalization* (Princeton University Press, 2003). But see Andrew Rose, "Do WTO Members have a More Liberal Trade Policy?" *NBER Working Paper* 9437 (2002).

3 On the latter see John Croome, *Reshaping the World Trading System: A History of the Uruguay Round* (The Hague: Kluwer Law International, 1999), 42–5 and 166–7.

4 There is also limited jurisprudence in oil and oil products. See Melaku Geboye Desta, "The GATT/WTO System and International Trade in Petroleum: An Overview," *Journal of Energy and Natural Resources Law* 21 (2003): 385–98; Melaku Geboye Desta, "The Organization of Petroleum Exporting Countries, the World Trade Organization, and Regional Trade Agreements," *Journal of World Trade* 37 (2003): 523–51; United Nations Conference on Trade and Development, *Trade Agreements, Petroleum and Energy Policies* (Geneva: United Nations, 2000).

Instead of GATT/WTO, the most important international institution facilitating cooperation between the major oil-importing states is the International Energy Agency (IEA). Its function is very different to trade facilitation, however. Instead, the IEA coordinates national policies on the drawdown of strategic petroleum reserves, and it has no authority as a platform for negotiating changes in the policies of member states. A secondary role of the IEA is providing information on energy markets globally, and peer reviews are carried out of member states' energy policies every few years as part of this process. These reports are for informational purposes only, however, and the recommendations contained in them do not bind the range of choices available to governments as they design and implement energy policy.

A second version of the transaction costs argument focuses on technological innovation. In financial markets in particular, technological change is proposed to limit the ability of governments to regulate capital flows. Digitalization has also enabled firms to relocate segments of the value chain internationally in ways that were not possible in previous decades. Both of these changes, it is argued, have increased the ability of firms to "exit" national economies, thereby increasing their bargaining power with governments.[5] In the case of oil, however, the industry has long been international in scope. Technological change has certainly had a profound effect on the oil industry. The growth in the size of tankers enables more than two million barrels to be transported in a single shipment, increasing the opportunities for international trade, and innovation in production enables firms to identify and reach oil fields at increasing depths and complexity through the use of 3D imaging, horizontal drilling, and other techniques. Many of these innovations were influenced by the nationalizations of oil production in the 1970s, however, which forced the oil majors to shift to areas such as the Gulf of Mexico and the North Sea in order to replenish reserves. In other words, they have helped firms to come home, rather than to leave.[6]

Another important explanation for the shift towards more liberal forms of market governance focuses on the role of economic power.

[5] John B. Goodman and Louis W. Pauly, "The Obsolescence of Capital Controls? Economic Management in an Age of Global Markets," *World Politics* 46 (1993): 50–82; Robert Solomon, *Money on the Move: The Revolution in International Finance Since 1980* (Princeton University Press, 1999), 110–11.

[6] See, for example, Tyler Priest, *The Offshore Imperative: Shell Oil's Search for Petroleum in Postwar America* (College Station, TX: Texas A&M Press, 2007).

Here, rather than governments using international organizations in order to achieve mutual gains, the decision to lower barriers to market entry occurs because of pressure from powerful states in the international system. In particular, the United States is argued to have played an important role in promoting trade and financial liberalization through a mixture of positive and negative sanctions.[7] One well-known example is the efforts of US officials to open Japanese markets to greater trade and investment in the 1980s and 1990s. In the case of oil, however, the United States has shown little capacity for coercing other countries in the international oil market, and has only sporadically attempted to pressure governments to open their energy markets.[8] Indeed, it is not clear that policymakers or firms in the United States are interested in opening trade in oil products in the first place.[9] Firms, for example, have often favored cartelization and protectionism rather than liberalization. The United States also implemented quotas on oil imports in the 1950s, and fixed prices in order to spur domestic production. Today the US continues to tax imports and limit exports of oil, albeit at a reduced amount, and billions of dollars of subsidies flow to firms operating in the United States in the name of promoting energy independence.

[7] The argument is developed in its general form by Stephen D. Krasner, "State Power and the Structure of International Trade," *World Politics* 28 (1976): 317–47. See also David A. Lake, "Leadership, Hegemony, and the International Economy: Naked Emperor or Tattered Monarch with Potential?" *International Studies Quarterly* 37 (1993): 459–89. For an application to financial liberalization see Eric Helleiner, *States and the Reemergence of Global Finance: From Bretton Woods to the 1990s* (Ithaca: Cornell University Press, 1994). Also see Daniel W. Drezner, *All Politics is Global: Explaining International Regulatory Regimes* (Princeton University Press, 2008).

[8] On the first point see Robert O. Keohane, "The Theory of Hegemonic Stability and Changes in International Economic Regimes, 1967–1977," in *Changes in the International System*, eds. Ole R. Holsti, Randolph Siverson, and Alexander George (Boulder, CO: Westview, 1980), 131–62; G. John Ikenberry, *Reasons of State: Oil Politics and the Capacities of American Government* (Ithaca: Cornell University Press, 1988). On the second, in the context of Japan, see Llewelyn Hughes, "Climate Converts: Institutional Redeployment, Industrial Policy, and Public Investment in Energy in Japan," *Journal of East Asian Studies* 12 (2012): 89–117.

[9] Llewelyn Hughes and Francisco Flores-Macias, "Drill Baby Drill! The Correlates of 'Energy Independence' Policy in the United States," Institute for International Economic Policy, George Washington University Working Paper IIEP-WP-2012–4 (2012).

An answer: firms and oil market governance

These are by no means the only explanations for economic liberalization. Economic crises, for example, have been identified as an important catalyst of economic liberalization.[10] In the case of oil the transformation of the structure of supply in the international market clearly forced both business and governments to reexamine whether the existing laws governing national oil market institutions continued to match their interests. In Chapter 3, and in the empirical chapters, I examine this, as well as considering how useful explanations for oil market liberalization that focus on the role of the European Union are, and the influence of ideas. But one of the interesting characteristics of oil is that we can discount the effects of some important explanations for economic liberalization when investigating changes in oil market governance.

How, then, are we to account for changes in oil market governance over the past three decades? In particular, how can we explain the shift to more liberal forms of economic governance in oil-importing countries, and ongoing intervention in support of domestic oil firms in some of them? In this book I develop an explanation that is rooted in the interests and strategies of domestic political actors. I am particularly interested in firms, and the way differences in firm characteristics affect the strategies they adopt in oil markets. I am also interested in where differences in firm characteristics come from. Indeed, a central contention of this book is that we need to trace the origins of firm characteristics if we are to understand how cross-national differences between oil firms emerged, and then show how this went on to shape their behavior.[11] The goal of the study is to go beyond simply identifying firm interests and strategies, however; I am also interested in explaining outcomes. I therefore link the strategies adopted by firms to policy outcomes by incorporating policymakers' incentives to intervene in oil markets.[12]

[10] Dani Rodrik, "Understanding Economic Policy Reform," *Journal of Economic Literature* 34 (1996): 9–41.

[11] Hall and Thelen point to the need to focus on firms as central actors in explaining economic liberalization. Peter A. Hall and Kathleen Thelen, "Institutional Change in Varieties of Capitalism," *Socio-Economic Review* 7 (2009): 7–34.

[12] On the "supply" and "demand" model sides of government intervention see Dani Rodrik, "Political Economy of Trade Policy," *Handbook of International Economics* 3 (1995): 1457–92.

In this book I am interested in the degree to which government instruments – in the form of laws, regulations, and policies – actively shape the behavior of market participants. A liberalized market lies at one end of a continuum, in which private actors interacting through the price mechanism play the predominant role in allocating goods and services.[13] Liberalization does not imply a simple reduction in the absolute size of government. It can be associated, for example, with a transformation of the role of the government towards monitoring and enforcing the terms of competition.[14] But in product markets it implies a standard set of policies, focused on freeing prices, reducing state monopolies through privatization, abolishing restrictions on trade, and imposing hard budget constraints on firms.[15] At the other end of the continuum we can define a market in which state-owned firms monopolize the production, importation, and distribution of oil and oil products.

One challenge in conceptualizing changes in the form of oil market governance lies in the role governments play as managers of risk.[16] Today, for example, the role of government as risk manager extends from regulating finance to protecting workers against income loss, accidents, or other losses of individual welfare. It also encompasses the energy sector. Natural resources differ from manufactured products because the location of production is determined by where they

[13] This is consistent with other discussions of economic liberalization. Laoyza and Soto, for example, refer to "the competitive participation of private agents in a sector, activity, or market." Norman V. Laoyza and Raimundo Soto, "On the Measurement of Market-oriented Reforms," *World Bank Research Working Paper WPS 3371* (2004), 3. Streeck and Thelen define liberalization as a "secular expansion of market relations inside and across the borders of national political-economic systems, significantly beyond the limits that the organized capitalism of the postwar 'mixed economy' had set for them." Wolfgang Streeck and Kathleen Thelen, *Beyond Continuity: Institutional Change in Advanced Political Economies* (Oxford University Press, 2005), 2. Simmons, Dobbin, and Garrett define liberalization as "policies that reduce government constraints on economic behavior and thereby promote economic exchange." Beth A. Simmons, Frank Dobbin, and Geoffrey Garrett, "Introduction: The International Diffusion of Liberalism," *International Organization* 60 (2006): 782–3.

[14] Steven Vogel, *Freer Markets, More Rules: Regulatory Reform in Advanced Industrial Countries* (Ithaca: Cornell University Press, 1996).

[15] Keith A. Darden, *Economic Liberalism and its Rivals: The Formation of International Institutions among the Post-Soviet States* (Cambridge University Press, 2009), 104–5.

[16] David A. Moss, *When All Else Fails: Government as the Ultimate Risk Manager* (Cambridge, MA: Harvard University Press, 2002).

are found, and the costs of extraction and transportation. Coal, which is the carbon residue of prehistoric plant life, for example, is widely distributed geographically. Oil on the other hand, which is the carbon residue of animal life, is distributed less broadly. This establishes an inescapable barrier to market entry: states and firms that control oil reserves can become producers; those that do not, cannot.[17] For much of its history, the central actors governing trade in oil globally were large, vertically integrated firms. These firms used a series of agreements to regulate the supply of oil and to distribute market share, and domestic oil markets were embedded in this international governance structure.[18] Business and governments then negotiated a series of what I call "industrial compacts" that regulated market share within national markets. Governments often also subsidized firms in order to increase their presence in exploration and production internationally. The promotion of the interests of national firms stands, I argue, at odds with the norm of non-discrimination that is at the heart of the world trading system.[19] Further, as I show in Chapter 2, governments across the advanced industrialized states abandoned this approach to oil market governance over the past three decades. I also show that, while every government reduced the extent of active intervention in oil markets, some wholly transformed their approach to managing risk in oil markets by abandoning the pursuit of greater control over the oil supply chain, while others adjusted existing institutions but retained vestiges of energy autarky. The goal of this book is to explain why.

Firms and oil market governance

National choices about how to manage risks in the international oil market are often analyzed through the "high politics" of statecraft. This makes some sense: oil remains enormously important to national economies and militaries, despite the replacement of some oil products

[17] Geology is an important determinant of the location and economic value of oil deposits, including rock porosity and the existence of geological formations that act as traps for the oil once it has formed. There are other theories of the formation of oil, most prominently abiogenic theory, which proposes that the origins of oil lie in inorganic compounds. For a review see Vaclav Smil, *Oil* (Oxford: Oneworld, 2008), 58–68.

[18] John M. Blair, *The Control of Oil* (New York: Vintage, 1978).

[19] Nowell points out that the bias towards domestic interests often also served the interests of international oil firms. Gregory P. Nowell, *Mercantile States and the World Oil Cartel, 1900–1939* (Ithaca: Cornell University Press, 1994).

with substitutes such as natural gas and nuclear fuels in electricity generation, heating, and some industrial processes. Oil is also a non-renewable resource, and the oil market is prone to bouts of instability. Both these factors have created uncertainty historically about whether supplies will be adequate to meet the core needs of the state.

The problem with framing policy outcomes in terms of statecraft, however, is that it ignores the fact that the most important entities that produce oil are firms, and that governments have imperfect control over their behavior. It also ignores the fact that many oil products are important inputs in productive processes in manufacturing and that gasoline prices are a pocketbook issue for voters. Recognizing this, I explain outcomes by focusing on the demands that firms make on governments, and the incentives governments have to meet these demands. I pay particular attention to how decisions to adopt more liberal forms of economic governance, as well as to retain energy autarky as a component of national policy, reflected the strategies of firms, and the way in which these strategies were shaped by their characteristics.

Rather than being motivated by national security, firm managers are concerned with profits. For most of the history of the oil industry it was accepted that profitability meant integrating operations vertically. This necessarily meant increasing firm size.[20] One reason this made sense is because the oil industry – like much of the energy sector – is capital intensive. This means that the unit costs of production tend to fall as the size of the firm increases. Another reason is because of the structure of competition. For the oil majors that emerged to dominate trade in oil and oil products, integrating vertically mattered because it enabled them to manage supplies globally; by regulating the flow of new sources of oil into national markets these firms sought to control prices and production. Indeed, prices remained remarkably stable up to and including the 1960s, despite the enormous increase in demand and supply globally.[21] For other firms, many of which began refining oil into kerosene, gasoline, and other products within domestic markets, vertical integration mattered because there were few centers of production open to investment outside the United States, and little

[20] For a discussion see Nick Antill and Robert Arnott, *Oil Company Crisis: Managing Structure, Profitability, and Growth* (Oxford: Oxford Institute of Energy, 2002). See also John G. McLean and Robert Haigh, *The Growth of Integrated Companies* (Norwood, MA: Plimpton Press, 1954).

[21] Blair, *The Control of Oil*, 100 and 117.

secondary market for oil and oil products. Under these conditions, integrating vertically meant gaining access to crude oil that weakened the control of the majors over them. The ownership of refining and marketing operations within domestic markets also promised these firms an avenue for disposing of oil produced.[22]

The emergence of an international market for crude oil, described in the next chapter, transformed the risks and opportunities facing firms in the long run by increasing the range of opportunities to buy and sell oil and oil products internationally. It also introduced the problem of price volatility. While this weakened the rationale for remaining vertically integrated on an operational basis, it did not wholly do away with the benefits of diversifying across business segments and countries as a strategy for managing price volatility and spreading risk. It also did not do away with the benefits of producing crude: the geological fact of the limited dispersion of oil internationally makes the structure of supply tend to oligopoly. This means that, over the long run, profits are higher in the production of oil than in refining, where the barriers to entry are lower. Moving into the production phase therefore means increasing profits.[23]

The ability to benefit from these new opportunities, however, as well as to manage the risks that came along with them, varied across firms. Some companies were able to exploit this new environment to their advantage without the need for support from government. Indeed, the existing rules governing oil markets at times hindered the ability of managers to improve their firm profitability. Other firms did not enjoy these advantages and were more likely to focus on non-market strategies centered on seeking the support of government as a result.

If firm characteristics explain the strategies they adopted, including the types of demands they made on governments, then how can we explain the origins of cross-national differences between firms? A common explanation for the competitiveness of industries cross-nationally, and by extension the types of policies they are likely to

[22] For a case study on Standard Oil (New Jersey) and efforts at integrating forwards into production see Alfred D. Chandler, Jr., *Strategy and Structure: Chapters in the History of the American Industrial Enterprise* (Cambridge, MA: MIT Press, 1969 [1962]), 163–224.

[23] An important exception is during the early period of domination by the Standard Oil Trust, which concentrated in transport and refining.

support, focuses on factor endowments.[24] When factors such as land, capital, or resources are relatively abundant, for example, then industries in countries that use the relevant factor intensively are expected to be more competitive. The distributional effects of market opening then enable us to infer the types of policies industries should support, with those that use an abundant factor intensively expected to gain from more open markets and therefore prefer liberalization, while those that use scarce factors intensively should favor greater protection.[25] We should also expect more internationalized industries to support liberalization over protection because of its implications for operational flexibility.

This approach offers a useful foundation for understanding how firms organize in oil markets. In terms of political organization, for example, the oil sector is characterized by high levels of asset specificity, and barriers to market entry are high. We should therefore expect firms to organize on a sectoral basis. The specificity of many of the assets employed in the oil industry also means that the returns to lobbying should be higher than in many other industries.[26] Yet, beyond this, there are complications with applying it to the problem of explaining the characteristics of firms in the oil industry and, by extension, the strategies they adopted towards government. This is because the majority of the member countries of the Organisation of Economic Co-operation and Development (OECD) produce almost no oil at all. Even in countries that do have oil reserves located within national borders, such as the United States, Norway, the United Kingdom, and Canada, the costs of production are high. Each is in a weak position, therefore, relative to Saudi Arabia and other major oil producers, on a global basis. Further, while internationalization matters, it is not clear why some oil firms succeeded in internationalizing, while others failed to do so.

[24] Standard models assume homogeneous firms within industries. Below I refer to cases in which factors of production are tied to their existing uses, a reasonable assumption in the oil industry.

[25] James E. Alt, Michael Gilligan, Dani Rodrik and Ronald Rogowski, "The Political Economy of International Trade: Enduring Puzzles and an Agenda for Inquiry," *Comparative Political Studies* 29 (1996): 689–717.

[26] On asset specificity see Jeffry Frieden, *Debt, Development, and Democracy* (Princeton University Press, 1991), 19–22; James E. Alt, Fredrik Carlsen, Per Heum, and Kåre Johansen, "Asset Specificity and the Political Behavior of Firms: Lobbying for Subsidies in Norway," *International Organization* 53 (2003): 99–116.

I argue that while managerial prowess and the timing of market entry undoubtedly affected the competitive position of firms, an important source of the differences between firms lies in a series of industrial compacts negotiated between business and government in national markets in earlier periods. These compacts crucially shaped the characteristics of firms and, by extension, the strategies they adopted in response to the transformation of the international oil market, including whether they chose to support or to oppose liberalization.

Rather than the reflecting the interests of firms or governments alone, these compacts were a bargained outcome between business and governments, and differed across countries. Sometimes these produced more diversified and larger firms. These firms were less likely to respond to the shift in opportunities and risks presented by the transformation of the international oil market described in the next chapter by seeking protection. Firms that took advantage of economies of scale were also less likely to demand support from the state, and indeed began to actively argue for liberalization when policies constrained profits. Firms that remained small and focused on single segments of the oil production chain, on the other hand, continued to demand government support. In this way the very success of these compacts in promoting the competitiveness of some firms led these firms to support, rather than oppose, the relaxation of barriers to international trade and investment.

The role of government

The importance of industrial compacts in shaping firm characteristics suggests that, rather than being captured by "Big Oil," policymakers have an incentive to shape oil markets that is independent from firm demands. Early interest in regulating oil markets often focused on limiting the corporate power of the international oil firms and their affiliates, which dominated the trade and distribution of oil and oil products in many national markets. There is also some evidence that oil policies were influenced by military planners, who were concerned about the supply of oil during wartime. Fears about oil scarcity were further stoked in the 1970s, when governments in the Middle East and North Africa nationalized control over a large share of global oil output.

While government interest in shaping oil markets may have a common origin in managing the market power of the international majors, however, a second interest lies in reducing the economic and political costs of oil price volatility. Support for oil firms operating in the oil industry is therefore balanced against the costs this imposes on firms that use oil products intensively, as well as individual consumers. Thus changes in oil market governance implemented over the past three decades are best understood as a bargained outcome produced by the strategies adopted by business and government.

In this book I place particular emphasis on how differences between firms shaped their approaches towards renegotiating the existing laws, regulations, and policies used to govern domestic markets. I show that differences in the industrial compacts negotiated between governments and business as they pursued these interests crucially shaped the characteristics of firms. It is these differences, I argue, that influenced their competitiveness, and hence how they responded to the emergence of an international oil market. When changes in the international oil market shifted the balance of costs and benefits associated with the status quo, firms that succeeded in increasing in scale and diversifying operations through earlier industrial compacts rejected them. Less competitive firms sought to adjust the existing institutions in the wake of the market volatility of the 1970s and 1980s, but did not seek to abandon them. Only by identifying the origins of firm heterogeneity, and how this affected the ways firms sought to renegotiate the institutional status quo, can we understand patterns of change in oil market governance.

In France, for example, business and government engaged in a complex series of negotiations over how to regulate the domestic market. The government was often divided and in a weak position in these negotiations. Yet they helped produce an equilibrium centered on first one, then two vertically integrated firms that enjoyed an international presence. The expanding interests of these firms outside national borders reduced the importance of the domestic market. It also conditioned their response to important changes in the international oil market that began to emerge in the 1960s. Indeed, when state instruments became a hindrance, as I show in Chapter 4, these firms actively lobbied to challenge the existing laws.

In Japan, on the other hand, the industrial compact negotiated between business and government included the erection of barriers to trade, and the provision of generous subsidies. These initiatives failed to promote the interests of firms in ways that enabled them to compete internationally, however. Instead, the structure of this intervention ensured that firms remained small on an international basis, and fragmented between the exploration and production phase, and refining and marketing within the domestic economy. Protection and support through the provision of capital therefore continued to be at the center of firm demands on government.

The case of the United States offers a useful contrast to that of Japan and France, allowing us to investigate how resource endowments and institutions combine to shape firm competitiveness. Oil production in the United States dominated the world oil market prior to World War II. One effect of this was to enable a small number of US firms to gain enormous market power internationally. These firms rarely turned to protectionism at home in order to support business goals, although they commonly supported barriers to entry in others' markets. Yet as Chapter 6 shows, the existence of domestic reserves of oil also allowed thousands of small firms, focused on producing oil within the United States, to flourish. As US production became uncompetitive relative to new producers in the Middle East and elsewhere, these firms turned to the government to protect market share.

The importance of oil market governance to the world economy

Changes in oil market governance matter enormously for the world economy. World spending on crude oil reached more than four percent of global GDP in 2007, and is estimated to remain at around five percent through 2030.[27] Cumulative investment in the oil sector is projected to reach almost six trillion dollars by 2030.[28] The billions of dollars spent in global trade and investment in oil and oil products mean choices about how to govern oil markets affect the distribution of wealth within and between states. One reason is that they shift economic rents from one producer country to another. For example, Chinese firms have invested substantial amounts in Angola, Gabon,

[27] On the value of spending on oil see International Energy Agency, *World Energy Outlook 2008* (Paris: OECD/IEA, 2008), 101–2. On the value of crude oil and oil products in world trade see Murray C. Kemp and Ngo Van Long, "The Role of Natural Resources in Trade Models," *Handbook of International Economics* 1 (1984): 367–417.

[28] Calculated in 2008 US dollars. International Energy Agency, *World Energy Outlook 2009* (Paris: OECD/IEA, 2009), 106–7.

and elsewhere in Africa, providing the capital to develop assets that had been relinquished by international oil companies.[29] These policies echo those adopted historically in Europe, Japan, and the United States, to improve the market positions of oil firms in upstream activities.

Intervention in oil markets also has distributive effects within oil-importing countries. One strategy used by firms and governments to manage risks in oil markets, for example, is to erect barriers to trade and investment in order to increase the market share of national oil companies, or to increase the oil produced domestically as a ratio of total oil consumed. Perhaps the best known case of the latter is the system of quotas and tariffs put in place by the Eisenhower administration. Established in the name of national security, it transferred five billion dollars in 1969 from consumers to producers, rising to eight billion dollars by 1980 in nominal terms.[30] Ongoing subsidies for domestic oil production confer billions of dollars on firms, and are justified as reducing the amount of "foreign oil" consumed domestically.

A second reason understanding patterns of state intervention in oil markets matters is because of international security. The *World Energy Outlook*, published annually by the IEA as the authoritative record of global supply and demand, notes that some sixty-four million barrels a day in new production is required to meet the projected growth in world demand and replace declining production from existing fields, approximately six times the current productive capacity of Saudi Arabia, the world's biggest producer of crude oil internationally.

An important feature of this new era is that the growth in global demand for oil comes from firms and consumers in China, India, and other emerging economies.[31] The problem of energy is therefore intertwined with a larger shift in the international balance of power. Some analysts have suggested that rising levels of imports in the new industrializing powers leads to an increase in resource nationalism as

[29] Erica S. Downs, "The Fact and Fiction of Sino-African Energy Relations," *China Security* 3 (2007): 42–68.

[30] Estimate by the Cabinet Task Force established by President Richard Nixon in 1969, which estimated consumer benefits caused by the removal of the oil import quota imposed to increase the competitiveness of oil produced domestically. For a summary see Douglas R. Bohi and Milton Russell, *Limiting Oil Imports: An Economic History and Analysis* (Baltimore, MD: Johns Hopkins University Press, 1978).

[31] International Energy Agency, *World Energy Outlook 2008*, 106.

these states seek to secure oil through national control. This is unlikely to lead to military conflict – wars over access to resources are rare. Nevertheless, the bargaining power afforded to states that produce oil can shape the foreign policies of oil-consuming states, as well as influence foreign policy decisions.[32]

Yet despite the importance of intervention in oil markets for the distribution of wealth within and between states, and its importance for international security, we have little systematic knowledge about how politics has shaped oil markets in the major oil-importing states over the past three decades, and how oil markets have shaped the politics of oil.[33] An additional goal of this book, therefore, is to contribute to our understanding of how importer states go about making decisions regarding how to manage risks associated with relying on the international markets for the supply of oil.

Overview of the book

In this book I make claims about the causes of changes in oil market governance in the advanced industrialized states over the past three decades. The core empirical section is made up of three country case studies. In Chapters 4 and 5 I examine changes in oil market governance in France and Japan. These cases were chosen because of substantial variation in firm characteristics, enabling me to examine how differences between firms shaped the demands places on governments by firms. They also enable me to hold differences in the position of the two countries in the oil market constant: both France and Japan have no reserves of oil within national borders, making differences in resource endowments an implausible explanation for differences in firm characteristics.[34] Although nuclear power plays a larger role in electricity

32 On the relationship between natural resources and civil war see Michael L. Ross, "What Do We Know about Natural Resources and Civil War?" *Journal of Peace Research* 41 (2004): 337–56.

33 Llewelyn Hughes and Phillip Lipscy, "The Politics of Energy," *Annual Review of Political Science* 16 (2013): 449–69. This contrasts with a growing body of research examining the politics of oil market governance in North Africa, the Middle East, Russia, and elsewhere, following the nationalization of oil-producing assets in many of these countries.

34 Differences in the presence of colonial possessions between France and Japan is also an implausible explanation for the variation in firm characteristics, as I discuss in the empirical chapters.

generation in France than Japan, both countries had substantial nuclear programs during the period covered in this study, and oil products continue to dominate in the transport sector, as indeed they do across the OECD countries.

In each chapter I trace how the industrial compacts negotiated between business and government shaped the characteristics of firms. In the case of France I also examine the role of Europe; as I outline in Chapter 3, the influence of European institutions offers an important alternative explanation for changes that have occurred since 1980. Data used to identify European policy preferences are drawn from primary sources obtained from the European Commission and Council of Ministers, in addition to interviews with firm and state actors, and secondary sources. Data on firm characteristics are drawn from annual reports and industry data. This is augmented by governmental and European assessments of the state of the industry where available. I use secondary literature to identify industrial strategies. There are a number of excellent studies of state policies in the oil sector that utilize primary source material for each of the cases. These sources are adequate to substantiate the claim that policies in earlier periods have shaped firm characteristics.

In the cases of France and Japan, I use hearings in which firms and industry associations make statements about their preferred policy outcomes to identify firms demands towards government. I also make use of data drawn from interviews with firm and industry association representatives. In order to identify what drove policymakers to shift their strategies towards oil market regulation, I conducted numerous interviews with participants in the setting of oil policies in Japan and France. Where possible I also interviewed representatives from the relevant political appointees and representatives from bureaucratic agencies in charge of decision-making at the time.

In Chapter 6 I extend the argument to the United States. Initial conditions in the United States differed markedly from both France and Japan. The case of the United States is useful because it allows us to investigate the relative importance of industrial compacts and resource endowments as factors influencing firm characteristics. The US case also enables us to examine how firm characteristics affected the strategies they adopted towards government unencumbered by differences in national institutions. This matters because while initial conditions in France and Japan were similar, and relations between firms and the government were close in both cases, a source of potential difference in the strategies adopted by firms that is not controlled for across these cases is the difference in political institutions between the two countries. In the case of the United States, on the other hand, we find substantial variation in firm characteristics within a single country. This allows an additional check on how firm characteristics affected the strategies they adopted towards government. In the case of the United States, as with France and Japan, I rely on annual reports and data produced by industrial associations and government bodies to examine firm characteristics. I link this to firm strategies using testimonies made by firm and industry representatives in congressional hearings. I also use congressional hearings, as well as interviews with officials, and media, to identify the positions of members of Congress and the executive towards intervention in oil markets.

The next two chapters lay the groundwork for the country cases that follow. In the first part of Chapter 2 I trace the evolution of intervention in oil markets in the advanced industrialized states. I do this to describe how governments developed a set of institutions designed to enhance national control over the oil supply chain as a response to oil market volatility. In the second part of the chapter I introduce an original index developed for the book in which I use data drawn from the International Energy Agency to examine changes in the institutions governing trade and investment in oil and oil products. I then use this index to examine the extent to which major oil-importing states have liberalized oil markets between 1980 and 2005.

In Chapter 3 I develop the explanation for changes in oil market governance in greater detail. I also identify two alternative explanations that could account for the changes outlined in these pages. While it is impossible to conduct an exhaustive test of their usefulness in explaining changes in oil market governance, I address them explicitly because of their plausibility. The first of these is rooted in ideas. While the risks of physical supply disruptions may have been real before World War II, when the international oil market was dominated by a small number of multinational oil firms and the regions that produced significant levels of oil were few, the increase in the distribution of oil reserves globally means there is now little risk that contracted supplies of oil will not be delivered. Given this, arm's-length trading represents an alternative, and more efficient, method for ensuring adequate supplies of oil are imported at lowest cost. Further, this structural shift also led to a change in understanding in how oil markets function, and therefore the risks associated with them. The influence of this change on policymakers' understanding of the risks in oil markets therefore represents a plausible alternative explanation, with cross-national differences driven by cross-national variation in the level of learning.

A second plausible explanation relates to the French case, and focuses on the role of Europe. The majority of the advanced industrialized states are located in Europe, and evidence suggests the European Commission actively sought to promote the liberalization of energy markets, including in oil. Thus it is plausible that the general changes in oil market governance are best explained by the influence of Europe on the national policies of a large subgroup of the advanced industrialized states.

In the final chapter I conclude by summarizing my findings. I also discuss the implications of these findings for our understanding of the governance of oil markets, economic liberalization more generally, and the rise of Asia-Pacific economies as important centers of demand for oil.

2 Oil markets and the physiocratic fallacy

Petroleum is an exhaustible natural resource; it is of fundamental importance to all phases of industrial activity and indispensable to industrial progress. At times it has been the subject of competition and rivalry; more frequently it has been the subject of agreement and international cartel arrangements.

The International Petroleum Cartel –
Staff Report to the Federal Trade Commission
August 2, 1952.

For much of its history the international oil market has been defined by the geographic distance between important oil-producing and oil-consuming regions.[1] The most important change in the organization of production, which gained momentum in the 1960s, was the renegotiation of the terms governing the production of oil from many of the world's most productive fields. As a result, the bulk of global oil reserves today are managed by national oil companies; firms such as Saudi Aramco today matter far more importantly to the global supply of oil than the Western oil companies that used to dominate the production of oil, although the increase in the production of unconventional oils in the United States and elsewhere promises to change this structure of supply once again. In the advanced industrialized states, on the other hand, I will show in this chapter that the most important change over the past thirty years has been the liberalization of domestic oil markets, although some countries continue to protect domestic interests in the name of securing oil.

In the first half of the chapter I briefly summarize changes in the structure of the demand and supply of oil and in oil market governance.

[1] J.E. Hartshorn, *Politics and World Oil Economics: An Account of the International Oil Industry in its Political Environment*, revised edn. (New York: Frederick A. Praeger, 1967), 45. The term "physiocratic fallacy" is drawn from Mancur Olson, Jr., "American Materials Policy and the Physiocratic Fallacy," *Orbis* 6 (1963): 670–88.

In the second part of the chapter I develop a strategy for conceptualizing and measuring changes in oil market governance across the major advanced industrialized states over the past three decades. I close the chapter by examining how the major oil-importing states changed the way they have governed oil markets over the same time period. This lays the foundations for the chapters that follow.

Changes in the international oil market

The total amount of crude oil within the earth's crust is both fixed and unknown. The composition of supply and demand for crude oil and products derived from it, on the other hand, as well as the organization of supply, has changed markedly over the past hundred years. These changes can be divided into three periods. In the first, the demand for products derived from oil shifted from lighting to become increasingly important in transport and industry, and therefore to militaries and civilian economies in the industrialized states.[2] Much of this growth in demand was met initially by firms based in the United States. Oil produced in the United States averaged more than sixty-five percent of total global production in years between the ends of World War I and World War II – followed by the Soviet Union (8.2 percent), Mexico (6.6), and Venezuela (5.1) – and US firms exported substantial volumes of oil and refined products. The formation of prices reflected this, with crude prices set by adding freight and insurance to Gulf of Mexico oil to establish the price for delivery at port.

Production from the Middle East, on the other hand, was not an important share of the world oil market, although business and government efforts to control reserves and manage oil production in the region were intense. Production from Iran, for example, averaged a little more than 130,000 barrels a day between 1918 and 1944, which represented 3.1 percent of world production, and production volumes in Iraq and Saudi Arabia were far smaller.[3]

[2] In the United States, for example, annual demand for kerosene increased from forty-three million barrels in 1918 to seventy-eight million barrels in 1944, while gasoline demand grew from eighty-five million barrels to 723 million barrels over the same period. Data here and paragraph below drawn from US Bureau of Mines, collected in DeGolyer and MacNaughton, *Twentieth Century Petroleum Statistics: Historical Data* (Texas: DeGolyer and MacNaughton, 2000).

[3] For simplicity I use Iran throughout the book, although its usage was formalized in 1935.

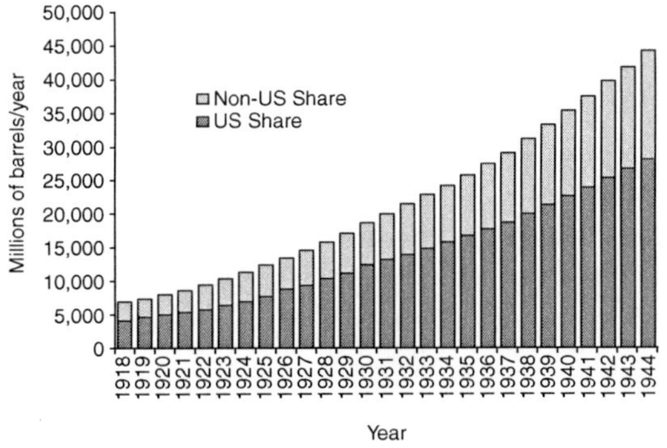

Figure 2.1 Crude oil production of the United States and the rest of the world (1918–44)
Source: Author calculations. Data drawn from DeGolyer and MacNaughton, *Twentieth Century Petroleum Statistics: Historical Data* (2000).

The second stage saw production diversify outside the United States. The dominant position of US oil was eroded by oil from the Middle East and elsewhere that was cheap to extract. The pace of the increase in both reserves and production from the Middle East, in particular, was astounding. A 1949 estimate put Middle East reserves at thirty-two billion barrels, compared to twenty-eight billion in the United States, yet from 1953 to 1957 proven reserves in the Middle East increased by twenty-three billion barrels annually.[4] Oil production from the Middle East also increased markedly, and was particularly important given that supply outstripped domestic demand; net exports from the Middle East increased from 671,000 to almost 1.2 million barrels a day between 1947 and 1949. The United States, on the other hand, became a net importer in 1948, and both exports of crude oil and oil products were negligible between the 1950s and 1970s.[5]

Across both these periods, the organization of crude oil production outside the United States, the Soviet Union, Canada, Mexico,

[4] Federal Trade Commission, *The International Petroleum Cartel – Staff Report to the Federal Trade Commission* (Washington, DC: United States Government Printing Office, 1952), 5–6; Francisco R. Parra, *Oil Politics: A Modern History of Petroleum* (London and New York: I.B. Tauris, 2004), 35.

[5] Federal Trade Commission, *The International Petroleum Cartel*, 12–13, and US Energy Information Administration data. US trade volumes were influenced by import barriers imposed in 1959. For details see Chapter 6.

and a few other producers, remained dominated by vertically integrated international oil companies.[6] An estimate by the Federal Trade Commission suggests these firms controlled approximately ninety-two percent of global oil reserves outside these countries in 1949, and eighty-eight percent of non-US and Soviet production.[7] These firms produced, processed, and delivered oil and oil products through intercompany transfers, and maintained a number of long-term contracts with one another. They used their financial power and technical expertise to establish long-term concessional contracts in many non-US oil-producing countries. Governments in these countries ceded important production and pricing decisions to them, and they commonly enjoyed concessions in excess of fifty years, while contract terms did not stipulate spending requirements on exploration, production, or reinvestment rates. Firms also made pricing decisions unilaterally.[8] In the most important consumer markets they often managed supplies through subsidiaries and affiliates specializing in the marketing and distribution of oil products, as well as selling oil to local refiners.[9] Oil market governance by the integrated oil majors also limited competition within and between the different segments of the supply chain.[10]

[6] The "majors" refers to Exxon (Standard of NJ), Shell, BP (British Petroleum, originally Burma Oil and Anglo-Iranian), Gulf, Texaco, Mobil (Standard of NY), Chevron, Standard Oil Co. (New Jersey), Standard Oil Company of California, Socony-Vacuum Co., Inc., Gulf Oil Corp., The Texas Co., Anglo-Iranian Oil Co., Ltd. (BP), and Royal Dutch Shell.

[7] The majors also controlled seventy-seven percent of non-US and Soviet refining capacity. Federal Trade Commission, *The International Petroleum Cartel*, 23–4. There were differences between the majors, with firms such as Royal-Dutch Shell more heavily invested in oil production, while the Standard Oil Company initially more concentrated in refining.

[8] The distribution of production rights varied between the firms. BP, for example, held the strongest position in Iran, with a lesser stake owned by Shell, while Esso, Socal, and Texaco dominated in Saudi Arabia. In Kuwait, on the other hand, BP and Gulf Oil were the most important partners. For production figures on a firm–country basis between 1950 and 1960 see Morris A. Adelman, *The World Petroleum Market* (Washington, DC: Resources for the Future, 1972), 80–1.

[9] See Federal Trade Commission, *The International Petroleum Cartel*, Charts 13–17 for the structure of marketing operations controlled by the majors.

[10] Paul Stevens (ed.), *Strategic Positioning in the Oil Industry: Trends and Options* (Abu Dhabi: The Emirates Center for Strategic Studies and Research, 1988), 8–9; Alberto Clo, *Oil Economics and Policy* (Dordrecht: Kluwer Academic Publishers, 2000), 68.

Emergence of a global oil market

The system of production and supply governed by the international majors ensured crude oil prices remained stable as long as the firms did not undercut each other in national markets and few new firms were able to enter the market. Crude oil prices varied less than five percent on an annual basis in twenty-two of the thirty years between 1940 and 1970, which is astonishing given the enormous growth in demand and supply.[11] Yet, in the long run, the firms were unable to deter new market entry. Maintaining control over production internationally proved impossible as supplies proliferated. One evaluation estimates the amount of crude oil traded at "arm's length" at 7.2 percent of the total market of traded oil globally, growing to 17.6 percent in 1957 and 20.7 percent in 1968. Another finds that the majors' share of production in regions other than the United States and the Soviet bloc fell from ninety-two percent in 1955 to seventy-six percent in 1965, and in refined products from eighty-one percent to fifty-eight percent over the same period.[12] By 1958 there were some 190 US companies, as well as non-US firms, carrying out exploration and production activities internationally, compared to twenty-eight before 1945.[13]

The inability of the majors to deter entry led governments in a number of oil-producing counties to organize against falling prices.[14] The Organization of the Petroleum Exporting Countries (OPEC) was formed in 1960 as price discounts became increasingly common and were offered on larger and larger volumes of crude for increasingly long periods. Governments in oil-producing countries gradually secured a larger share of profits, before increasingly taking control over production decisions.[15] Governments often established national oil companies to participate in the exploitation of

[11] Clo, *Oil Economics and Policy*, 63.

[12] Steven A. Schneider, *The Oil Price Revolution* (Baltimore, MD: Johns Hopkins University Press, 1983), 86. See also Parra, *Oil Politics*, 67.

[13] Helmut J. Frank, *Crude Oil Prices in the Middle East: A Study in Oligopolistic Price Behavior* (New York: Frederick A. Praeger, 1966), 95.

[14] Schneider points out that quotas placed on imports to the United States also placed downwards pressure on global prices. Schneider, *The Oil Price Revolution*, 46.

[15] The best account of the formation and early years of OPEC is in Parra, *Oil Politics*. Major nationalizations of concession-holder operations took place in Algeria and Libya (1971), Iraq (1972–5), Libya once again (1973), and Kuwait, Qatar, and Venezuela (1976).

resources, including Sonatrach in Algeria in 1963, the Iraqi National Oil Corporation in 1964, PDVSA in Venezuela in 1975, the Kuwait Petroleum Corporation in 1976, and the Nigerian National Petroleum Corporation in 1977.

The transfer of production and pricing decisions to governments in the Middle East, North Africa, and elsewhere, did not do away with the problem of market power. It did, however, transform the structure of the oil market from one of monopolistic stability to one characterized by competitive instability.[16] Instead of being dominated by the vertically integrated majors – which internalized decisions over the production, refining, and distribution for much of the global oil demand – the management of prices and production shifted to governments in the major oil-producing countries, and particularly Saudi Arabia, although the degree of market power they enjoyed was far less than that enjoyed by the majors.[17]

The fragmentation of the organization of the oil market meant crude was increasingly traded in an external market in which buyers and sellers engaged in arm's-length contracting on a long-term or spot basis. Prices also began to be set through index-based pricing rather than being administered by the majors or OPEC.[18] Data on the structures and volumes of oil contracts globally remains limited, but one estimate finds that oil traded at arm's length, measured by the difference between crude oil produced and refinery capacity held by the top fifty oil producing firms, reached almost twenty-two million barrels a

[16] These terms are taken from Clo, *Oil Economics and Policy*.

[17] There remains disagreement over the extent to which OPEC, and Saudi Arabia, are able to effectively control oil prices through adjusting levels of production. A number of models of market power have been developed and tested, from the cartel model focused on OPEC, to a dominant producer model focused on Saudi Arabia. See, inter alia, Dermot Gately, "A Ten-Year Retrospective: OPEC and the World Oil Market," *Journal of Economic Literature* 22 (1984): 1100–14; Carol Dahl and Mine Yücel, "Testing Alternative Hypotheses of Oil Producer Behavior," *The Energy Journal* 12 (1991): 117–38; Robert K. Kaufmann, Stephane Dees, Pavlos Karadeloglou, and Marcelo Sanchez, "Does OPEC Matter? An Econometric Analysis of Oil Prices," *The Energy Journal* 25 (2004): 67–90; James L. Smith, "Inscrutable OPEC? Behavioral Tests of the Cartel Hypothesis," *The Energy Journal* 26 (2005): 51–83.

[18] Bassam Fattouh, "An Anatomy of the Crude Oil Pricing System," *Oxford Institute for Energy Studies WPM* 40 (January 2011). Market-based pricing was first adopted by PEMEX in 1986.

day by 1995.[19] This transformation of the oil market also increased long-run and short-run price volatility, compared to the period of the majors' dominance. Between 1930 and 1970 the real crude oil price (US$2011) varied between nine and twenty dollars on an annual basis, with a mean price of fifteen dollars. Between 1973 and 2011, on the other hand, prices varied between sixteen and 111 dollars on an annual basis, with a mean of fifty dollars a barrel. This included the peaks of 1973–4 and 1978–9, as well as the collapse in prices in the 1980s, when the price fell to thirty dollars in 1986, and eighteen dollars in 1998 in real terms.[20]

Diversification in oil supplies through the development of new sources of oil was one cause of the fall in prices.[21] Exports of Mexican crude, for example, jumped from a little over 500,000 barrels per day in 1979 to a peak of 1.5 million barrels in 1984.[22] Production in the North Sea also grew. In the UK-controlled area it jumped from an average of 1.67 million to 2.68 million barrels per day between 1980 and 1986, and Norwegian production increased from an average of 528,000 to 806,000 barrels per day over the same period. Firms also won market share by undercutting the official price; OPEC's share of global oil production fell to twenty-nine percent in 1986 from a peak of fifty-two percent in 1973.[23]

Production growth outside OPEC confounded predictions it would be insufficient to meet growing demand. The number of countries producing more than 100,000 barrels per day increased from thirty-six at the end of the 1970s to forty-four in 1990, and forty-six in 2000. In Latin America between 1990 and 2000, Brazil, Columbia, Ecuador, Argentina, and Mexico added at least 100,000 barrels per day, while

[19] Giacomo Luciani and Mario Salustri, "Vertical Integration as a Strategy for Oil Security" in Stevens (ed.), *Strategic Positioning in the Oil Industry*, 23–44.

[20] Data from BP Statistical Review of World Energy 2011. Prices to 1944 are the US average, Arabian Light posted at Ras Tanura between 1945 and 1983, and Brent dated between 1984 and 2011.

[21] Fadhil J. Al-Chalabi, *The World Oil Price Collapse of 1986: Causes and Implications for the Future of OPEC* (Washington, DC: Johns Hopkins University Press, 1987), 131–7. Most analyses also point to falling demand and changing OPEC pricing policies.

[22] George W. Grayson, *Oil and Mexican Foreign Policy* (University of Pittsburgh Press, 1988), 41.

[23] Hillard G. Huntington, "Oil Price Forecasting in the 1980s: What Went Wrong?" *The Energy Journal* 15 (1994): 12.

in Asia-Australasia firms in Australia, China, Malaysia, India, and Vietnam added at least 100,000 barrels per day to production. African supply also increased: Algeria, Angola, Egypt, Nigeria, and the Sudan added at least 100,000 barrels per day to production over the same period.[24]

Comparatively low prices for oil ended in the 2000s. Prices then appeared stable until the latter half of 2003, when they increased dramatically, breaking seventy dollars a barrel at the end of 2006, and increasing to over 140 dollars a barrel in 2008 in nominal terms. The explanation for this rise in prices in the 2000s is contested. An Interagency Task Force's Interim Report on Crude Oil published in July 2008 found that the balance of supply and demand best explained the increase in crude oil prices, noting that global oil consumption grew 3.9 percent over the 2004–7 period, while surplus capacity was lower than historic levels.[25] In an alternative view, financial speculation distorted futures markets for oil contracts, with this distortion transmitted to the spot market that determines the prices for crude oil and oil products such as gasoline that are traded physically. The rise of production in the United States today after decades of decline also signals a new and important change in the structure of global oil supply. Regardless, price volatility is now a structural feature of the international oil market, when compared to the stability of the majors-dominated production system of the 1950s.

Oil market governance in the advanced industrialized states

The growth in demand, lack of substitutes, and dominant position of the majors, made the organization of the production of oil and oil products important to governments. For firms, on the other hand, the organization of supply for oil and oil products mattered for profits.

[24] The exception to this steady growth in production in the non-OPEC areas is the former Soviet Union, where it declined precipitously following its collapse; production from the Russian Federation, Kazakhstan, and Azerbaijan fell from 11.2 to 7.2 million barrels per day between 1990 and 1999. The most significant producers in the former Soviet Union are the Russian Federation, Kazakhstan, Azerbaijan, and Uzbekistan. All figures calculated by author from BP Statistical Review of Energy.

[25] Commodity Futures Trading Commission, *US Interagency Taskforce on Commodities Markets, Interim Report on Crude Oil* (Washington, DC: CFTC, 2008).

In response, government and firms across the advanced industrialized states negotiated a series of industrial compacts that regulated the distribution of market share within national markets, and influenced access to the production of oil internationally. Often, one component of these industrial compacts was the creation of new firms as a vehicle for business–government cooperation. In France, for example, laws designed to regulate market shares emerged after a long period of competition for control over the supply and distribution of oil, and the profits derived from it, within the French market. A complex series of negotiations ended with the creation of the Compagnie Française des Pétroles (CFP) and Compagnie Française de Raffinage (CFR), which helped stabilize the distribution of market share between majors and domestic firms. The Bureau de recherches de pétrole (BRP) was established in October 1945. Unlike the CFP it was wholly government-owned, and focused on exploiting gas within France and oil discovered in the Franc Zone in North Africa and elsewhere. What was to become the Elf-ERAP group was granted a twenty-five percent share of the domestic refining market by law through the Union générale des pétroles (UGP).

In Italy, the national oil firm Azienda Generale Italiano Petroli (AGIP) was founded in 1926 with the goal of breaking the duopoly of the majors.[26] Gas and some oil were found domestically in the Po Valley after World War II. A 1953 law established a state monopoly over oil and gas exploration and production domestically, and Ente Nazionale Idrocarburi (ENI) was established as a holding company to centralize state companies involved in the oil and gas sectors, and expanded its activities into Iran and Egypt through production sharing agreements. By 1971 ENI controlled twenty-five percent of the domestic market and was almost self-sufficient in oil on a firm basis.[27]

Market governance in Japan was conditioned by an almost complete lack of domestic reserves of petroleum, and the control by majors' subsidiaries of the domestic market. The first industry-wide legislation gave the government the authority to control trade flows and refining capacity through licensing, as well as to control prices and distribute

[26] Pier Angelo Toninelli, "Energy Supply and Economic Development in Italy: The Role of the State-Owned Companies," *Department of Economics Working Paper Series No. 146* (October 2008); Leslie E. Grayson, *National Oil Companies* (Chichester: John Wiley and Sons, 1981).

[27] Grayson, *National Oil Companies*, 113.

market share between domestic refiners and marketers. A public corporation was also established to provide fiscal support, in the form of subsidies and low-interest loans, to Japanese private sector firms operating in exploration and production internationally.

Nationalizations and price volatility in the 1970s appeared to herald a new era of energy insecurity. Governments formulated diplomatic strategies in response to the crisis, and promoted investment in partial substitutes for oil products such as nuclear power, natural gas, and coal. Fuel diversification and energy efficiency were codified in the Shared Goals agreed upon by the member states of the International Energy Agency (IEA), and adopted by that body in 1993. But what of the institutions governing national oil markets? In the next section I conceptualize and measure changes in oil market governance. I show that the advanced industrialized states made important changes to the domestic laws used to govern oil markets. Most notably, although all the major oil-importing states had substantially liberalized oil markets since 1980, there also existed important variation in the degree to which these states liberalized their oil markets.

Defining change in oil market governance

In this section I focus on two instruments used to regulate the market share of firms domestically and internationally: barriers to trade in oil and oil products, and state ownership of oil companies. Both forms of state intervention are, I argue, incompatible with liberal conceptions of economic governance. Evidence to support this approach can be found in the IEA Shared Goals, which codify that energy security policies should be guided by the principles of diversity, efficiency, and flexibility. National policies, according to the Shared Goals, should promote the efficient use of energy across all stages of energy production and consumption, including through the research, development, and deployment of technologies that improve energy efficiency. Policies should also promote diversification of energy sources, particularly towards non-fossil fuels.

The Shared Goals imply the rejection, on the other hand, of tariffs and quotas, the provision of capital to national oil firms, or other measures, noting that "free and open trade and a secure framework for investment contribute to efficient energy markets and energy security. Distortions to energy trade and investment should be avoided."[28] This is unsurprising. Allocating market share through quotas, tariffs, licenses, or other forms of regulation stands in opposition to the postwar trade regime that, as a political project, was designed to bind states to policies that reject nationalist forms of economic governance. The policy instruments used to intervene in the oil market commonly include ownership of national oil firms and/or the use of trade, fiscal, and regulatory policies to regulate market share between firms, as well as to subsidize the production of oil within national boundaries – where it exists – or the production of oil products through refining. Given this, more liberal forms of governance in oil markets should be associated with the abolition of laws and regulations discriminating in favor of particular firms, or the production of oil or refining of oil products domestically.

Measuring change in oil market governance

One plausible way to measure changes in oil market governance is through examining changes in government spending over time. A problem with this approach, however, is that expenditure levels vary according to factors unrelated to changing forms of governance, such as oil price volatility. This also misses forms of market governance that do not involve loan provision, subsidies, or other monetary transfers. An alternative strategy partially resolves this problem by attaching a monetary value to government intervention. The OECD, for example, measures agricultural subsidies this way, and it has also been adapted to measure levels of state support to the coal sector.[29] Attaching a value to state intervention in oil markets is challenging, however, given the wide variety of non-tariff measures used to shape outcomes in oil markets. Most importantly, forms of governance focus not only on the provision of subsidies, but extend to the ownership of national-flag oil companies, for which there is no obvious way of establishing a measure of levels of state support in commensurate terms.

A different strategy is to focus directly on changes in laws, regulations, and policies. Here I measure changes using an index that tracks

[28] Shared Goals of member states of the International Energy Agency. Available at www.iea.org (accessed August 7, 2010).

[29] Kristi Varangu, "IEA Work on Defining and Measuring Environmentally Harmful Subsidies in the Energy Sector," *OECD Workshop on Environmentally Harmful Subsidies* (Paris: OECD, 2002).

the presence of laws, regulations, and policies designed to affect the market share of firms involved in the production and refining of oil. I examine changes in oil market governance for countries that have negligible domestic resources of oil, defined as those that rely on imports for more than ninety percent of total domestic consumption over the period examined in the study.[30] This includes the majority of the advanced industrialized states; only Australia, Canada, Denmark, Mexico, Norway, the United States, and the United Kingdom have domestic supplies of oil.[31]

The index was created using data on national policies drawn from the IEA annual publication *Energy Policies and Programs of IEA Countries*, for the years 1978–2006. These were augmented by individual country reports produced by the IEA for member countries every four years, beginning in 1996, as well as secondary materials that offer further detail on the institutions in individual countries governing oil markets.[32] Using this information, I have coded legal and regulatory intervention for two categories: state ownership and trade. Policies are coded by full, partial, or no liberalization, as described below.[33]

30 I limit myself to these countries because the logic of oil market governance in resource-abundant countries can differ in oil-producing states. For a review see Pauline Jones Luong and Erika Weinthal, "Rethinking the Resource Curse: Ownership Structure, Institutional Capacity, and Domestic Constraints," *Annual Review of Political Science* 9 (2006): 241–63.

31 South Korea and Iceland are not included in the index for lack of data. The former Soviet states are also omitted. Luxembourg is not included in the index as it has no refinery facilities.

32 For example, Massimo Nicolazzi, "Petroleum Law and Petroleum Joint Ventures in Italy," *Journal of Energy and Natural Resources Law* 8 (1990): 30–50; Michael O'Connor and James Bruining, "Regulation of Offshore Petroleum Activities in Ireland," *Journal of Energy and Natural Resources Law* 19 (2001): 132–42; Ilias Bantekas, "Hellenic (Greek) Petroleum Law: Regulation of Exploration and Exploitation," *Journal of Energy and Natural Resources Law* 12 (2002): 117–29; Selma Stern, "Turkey's Draft Petroleum Law," *Journal of Energy and Natural Resources Law* 23 (2005): 308–22.

33 Adapted from International Energy Agency, *The Role of IEA Governments in Energy: 1996 Update* (Paris: International Energy Agency, 1996): 24. Coding methodology is adapted from Graciela Kaminsky and Sergio Schmukler, "Short-Run Pain, Long-Run Gain: Financial Liberalization and Stock Market Cycles," *Review of Finance* 12 (2008): 253–92. Policy instruments mirror those identified in Grayson's cross-national study of national oil companies in oil-importing states. See Grayson, *National Oil Companies*, 234–42.

State ownership: government shareholding in national oil firms is one component of oil market governance. The provision of capital through equity participation in oil firms was a common feature of oil market governance even in those states that have little domestic oil. This does not imply governments have perfect control over firms. The interests of domestic firms and governments are often aligned, but this is not always the case. Even in oil-producing countries, firms differ in the extent to which ownership confers control.[34] I code an outcome as having a high level of liberalization when the state has no ownership stake in an oil company operating in the oil sector, on the principle that no state ownership offers less control than in the case that the government retains state ownership. This includes relinquishing ownership of a "golden share" giving it special rights to the government. Outcomes are coded as partial liberalization when the government holds a minority share, including maintaining a golden share. The oil sector is coded as having a low level of liberalization when the state owns a majority share, or completely owns a firm operating in the oil sector.

Trade: abolishing barriers to trade and investment between national economies lies at the core of the postwar project of promoting economic growth. Trade in crude oil and oil products was commonly subject to both tariff and non-tariff measures that redistributed market share. Trade barriers were used to encourage the construction of refineries within national borders in Japan and France, for example, by discriminating against oil refined internationally. Import licenses and other measures also allocated market share for oil products.

A country is coded as having a high degree of liberalization when the import or export of oil or oil products is regulated on a non-discriminatory basis. This implies the absence of quotas or tariffs applied to the import or export of oil or oil products in order to increase domestic production of crude oil, or refined products, or to increase the market share of domestic firms. A country is coded as having partially liberalized institutions of

[34] Luong and Weinthal, "Rethinking the Resource Curse." One interviewee identified important differences between CFP and Elf Aquitaine in the tightness of firm–government relations. Gilles Bellec, Director of Hydrocarbons (1984–9), Ministry of Industry, Paris, France, interview with author, July 18, 2008.

governance over oil markets when domestic firms are guaranteed a share of imports, but do not monopolize trade in oil or oil products, or alternatively when quotas or tariffs are placed on the import or export of crude oil or oil products in order to increase domestic production. Finally, a country is coded as having a low level of liberalization when the import or export of oil or oil products is banned, or alternatively when the trade in oil or oil products is monopolized by national-flag firms.

A table summarizing this coding scheme is shown below.

Changes in oil market governance in the major oil-importing states 1980–2005

The reports produced by the IEA do not standardize the policy instruments reported on and are an imperfect source for cataloguing the changes in oil market governance across the fifteen countries that are included in the analysis. They do represent a credible source of data, however, from which it is possible to draw information on the types of policies implemented in oil markets on a cross-national basis. I present the results in Figure 2.2 below for the fifteen countries included in the analysis. Where there has been little liberalization, the country is given a score of zero in the relevant category; where there is partial liberalization the country is given a score of one; finally, where there is a high level of liberalization the country is given a score of two.

The first thing to note is that the index only measures changes in barriers to trade and in the provision of capital to oil firms through equity ownership. While these are important components of the institutions governing oil markets, price controls, subsidies, and other policies have also been used to shape oil markets. I take up discussion of the wider set of instruments in the empirical chapters that follow.

With this caveat, what does the data show us?

Most obviously, the mean direction of institutional change since 1980 is in favor of more liberal forms of economic governance. In almost all fifteen countries surveyed, governments reduced the use of trade-based policy instruments designed to shape flow of oil or oil products, and divested stakes held by states in national oil companies. The exceptions are Belgium and the Netherlands, which began and ended the period with high levels of liberalization across both

Table 2.1 *Measuring oil market governance*

State Ownership	
Criteria used for Low Liberalization (0 points)	
Ownership	Mention of full ownership of firm(s) by state or state owns majority shareholding in firm
Partial Liberalization (1 point)	
Ownership	Mention of state ownership of firm(s) in all sector; state owns minority shareholding in firm and/or retains "golden share"
High Liberalization (2 points)	
Ownership	No mention of state ownership of firm(s) in oil sector
Trade	
Criteria used for Low Liberalization (0 points)	
Import/export controls	Mention that import or export of oil or oil products banned; mention of import or export monopoly by domestic firm(s)
Partial Liberalization (1 point)	
Import/export controls	Mention of tariff or non-tariff measure on import or export of oil or oil products; mention of domestic firm(s) guaranteed share of imports/exports
High Liberalization (2 points)	
Import/export controls	No mention of tariff or non-tariff measure on import or export of crude oil or oil products; no mention of allocation of share of imports/exports to domestic firm(s)

categories. There were no countries that reversed this overall trend towards greater liberalization.

Second, while the mean direction of change was towards reducing intervention in oil markets since 1980, there was also variation between countries: some wholly liberalized oil markets were transforming the strategy used to manage risk in the international oil market,

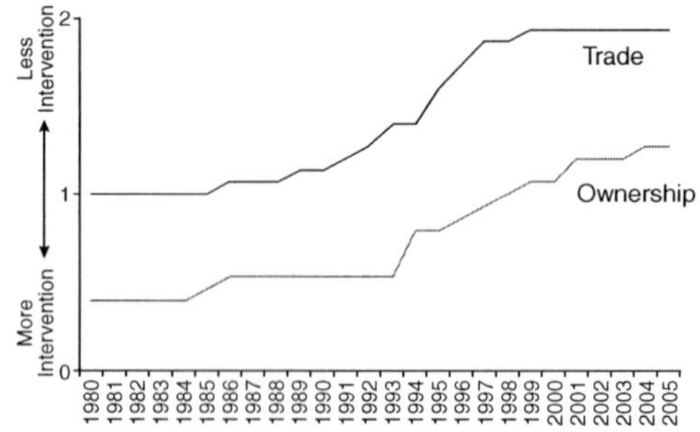

Figure 2.2 Mean change in oil market governance (1980–2005)
Source: Data drawn from International Energy Agency, *Energy Policies and Programs of IEA Countries*, various years; International Energy, *Energy Policies of IEA Countries*, various years.
Note: Includes Austria, Belgium, Finland, France, Greece, Ireland, Italy, the Netherlands, Portugal, Spain, Sweden, Switzerland, Turkey, Japan. Luxembourg omitted due to country size.

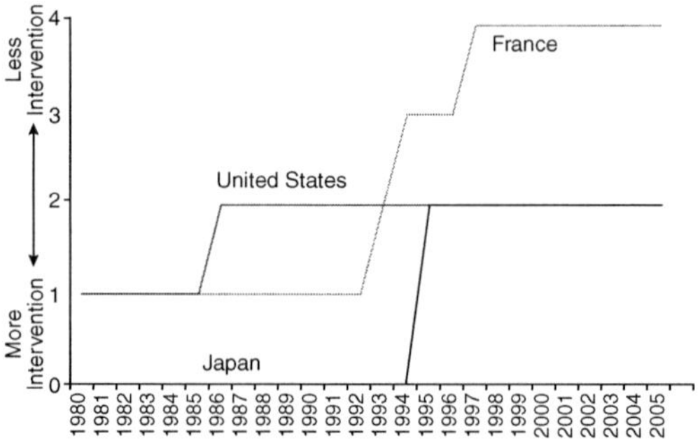

Figure 2.3 Changes in oil market governance in France, Japan, and the United States (1980–2005)
Source: Data drawn from International Energy Agency, *Energy Policies and Programs of IEA Countries*, various years; International Energy, *Energy Policies of IEA Countries*, various years.
Note: Ownership and Trade indices combined to produce 4-point scale.

while others continued to intervene in oil markets through the use of trade measures or ongoing ownership of state firms. The French and Japanese cases examined in this book and shown in Figure 2.3 above, for example, began with low levels of liberalization in the oil market across the two categories, before moving to higher levels of liberalization. This was also the case with the United States for which the data has also been added above. While France fully liberalized, however, Japan and the United States did not. It is these changes that I seek to explain in the chapters that follow.

3 *Explaining changes in oil market governance*

In the last chapter I showed that governments made important changes to the way they governed oil markets over the past three decades. Laws and regulations used to govern domestic oil markets were largely replaced by strategic petroleum reserves coordinated through the International Energy Agency (IEA), with the distribution of oil otherwise left to the market except for regulations controlling product quality. I also showed that some governments retained components of the previous form of oil market governance, including their continued protection of the position of domestic firms and international promotion of their competitiveness.

In this chapter I develop an explanation for these outcomes. I pay particular attention to the strategies of firms, and make two claims. First, I propose that the strategies of firms had an important effect on the adoption of more liberal forms of oil market governance in national markets, as well as the ongoing active role of governments in allocating market share. Larger and more diversified oil firms became less interested in the resources available from the state, as opportunities proliferated along with the emergence of an international market for oil. This made them both less willing to invest resources in defending the status quo, and more willing to lobby government to change the status quo where the net benefit to doing so helped profitability. Smaller and less diversified firms, on the other hand, continued to seek support from government. Second, I propose that differences in the characteristics of firms cross-nationally, and by extension the types of policies they demanded from governments, reflected the effects of industrial compacts negotiated in earlier periods. Institutional settlements negotiated historically therefore mattered, shaping firm characteristics in ways that conditioned their responses to changes in risks and opportunities in the international oil market.

I begin the chapter by examining the explanations for economic liberalization, contrasting those that focus on international influences with those focusing on the interests of domestic actors. I also link this discussion with existing explanations for patterns of oil market governance in the advanced industrialized states.

I then contrast these with my own explanation, and discuss two plausible alternative explanations for outcomes. Oil markets offer an interesting contrast to many other areas of the economy because some important explanations we have for liberalization are irrelevant to oil. This allows us to discount them in seeking to understand changes in oil market governance. There are nevertheless two important alternative accounts for the changes documented in the last chapter, focused on ideational change and the role of Europe. I conclude the chapter by suggesting how we can adjudicate the relative importance of these and my own explanation.

Explaining changes in oil market governance

Explaining the shift towards more liberal forms of economic governance is a central area of inquiry in political economy. This is partly because of its enormous importance in redistributing wealth within and between national economies. It is also because it is puzzling. Standard accounts of government intervention in economic markets point to a status quo bias in policies. They also tend to assume that protectionism is overprovided relative to what is socially optimal because the gains from protection are concentrated in particular firms or industries, while the benefits of liberalization are diffuse. This leaves us with a real puzzle: over the past thirty years we have seen tremendous institutional changes, and these changes have not been in the direction of greater protectionism. Instead, governments favored liberalization and privatization rather than erecting and maintaining barriers to market entry.

In the last chapter I showed that liberalization occurred in oil markets across the major advanced industrialized states. I also showed there continue to be important cross-national differences in how oil markets are governed. Why is it that barriers to trade and investment have fallen in oil markets across the advanced industrialized countries? And why is it that barriers have fallen, but some countries continue to intervene in favor of domestic firms? Below I consider a number of common explanations for economic liberalization and examine how useful they are for explaining changes in oil markets governance.

Transaction costs, power, and oil markets

The shift towards more liberal forms of economic governance is unlikely to be explained by any single, tractable model. Oil nevertheless represents an interesting case for exploring its causes because it is insulated from a number of important explanations for liberalization.[1] One prominent explanation for the liberalization of trade and capital flows, for example, focuses on falling transaction costs. International organizations, in particular, are argued to promote lower barriers to trade and investment by improving information flows between countries, and by providing certainty that trading partners will honor their commitments. International organizations also enable governments to link policies across different sectors, balancing gains with losses in ways that enable agreements to be reached that would otherwise remain out of reach.[2]

In the case of oil, however, international organizations have played only a limited role in promoting the reduction of barriers to trade and investment.[3] One explanation for the exception granted to oil is that it reflected a "gentleman's agreement" between governments in the United States and Europe, which sought to avoid politicizing the control of many of the world's most productive fields by the oil majors. The lack of production in many of the founding members of GATT meant there were few countries with strong exporter interests seeking to have oil and oil products included in negotiating rounds.[4] As a result, efforts to subject oil to multilateral rules governing trade and investment have been sporadic. The most important effort to include natural resources, including oil, in the multilateral rules governing trade and investment occurred during the Uruguay Round. The Punta del Este Declaration that launched the trade round in 1986 included a commitment to "achieve the fullest liberalization of trade in natural resource-based

[1] On liberalization as a complex phenomenon see James E. Alt, Jeffry Frieden, Michael Gilligan, Dani Rodrik, and Ronald Rogowski, "The Political Economy of International Trade: Enduring Puzzles and an Agenda for Inquiry," *Comparative Political Studies* 29 (1996): 689–717.

[2] Christina L. Davis, "International Institutions and Issue Linkage: Building Support for Agricultural Trade Liberalization," *The American Political Science Review* 98 (2004): 153–69.

[3] I discuss the European Union later in the chapter.

[4] See United Nations Conference on Trade and Development, *Trade Agreements, Petroleum and Energy Policies* (Geneva: United Nations, 2000), 14–15.

products," and the Negotiating Group on Natural Resource-Based Products established during the round took up the question of how to reduce or eliminate tariff barriers and non-tariff measures such as price fixing, subsidies, import and export restrictions, and government ownership of natural resource products. Disagreement between delegates over the scope of the negotiations, particularly over whether energy and energy-based products should be included in addition to non-ferrous metals and minerals, forestry products, and fish and fisheries products, remained unresolved, and there was little progress in including oil and oil products in the trade round.[5] The "essential security interests" clause included in Article XXI of the GATT is understood to apply to trade in oil and oil products, providing governments with a broad mandate to restrict trade and investment on national security grounds.[6]

Rather than the trade regime, the most important organizations promoting international cooperation in the oil market are reflections of the problem of market power identified in Chapter 2. On the producer side, the most important international organization is the Organization of the Petroleum Exporting Countries (OPEC), which was formed in 1960 to represent the interest of a number of global oil-producing countries in maintaining a floor on crude oil prices. On the consumer side, the most important international organization is the International Energy Agency (IEA), which was formed in 1974 in response to price volatility and the nationalizations of the 1970s. Today, the most important role of the IEA is to coordinate the emergency release of oil stocks held by member countries. The IEA also analyzes market conditions and facilitates the peer review of members' energy policies. These reviews are not binding, however, and the organization has no authority as a body through which member states can negotiate changes to policies used to govern energy markets on a national basis.

A second way in which falling transaction costs have contributed to economic liberalization is through lower transport costs, improvements

[5] GATT Secretariat, "Note by the Secretariat: Summary of Statements and Proposals Made Concerning Negotiations on Natural Resource-Based Products," Negotiating Group on Natural Resource-Based Products MTN. GNG/WG3/W/8/Rev.1, January 14, 1988.

[6] Donald N. Zillman, "Energy Trade and the National Security Exception to the GATT," *Journal of Energy and Natural Resources Law* 12 (1994): 117–27.

in communications technologies, and other changes that have eased the ability of firms to trade across national borders and increased the range of investment opportunities available to them.[7] The oil industry has certainly not been insulated from the effects of some of these changes. The deepening of capital markets, for example, has increased the range of options available to oil firms as they seek to raise capital for exploring for oil and for producing it. In other ways, however, falling transaction costs have been less important to the oil industry. In particular, while the industry is extraordinarily innovative, new technologies have tended to enable firms to stay at home, rather than increase their opportunities for exit. They focus on reducing the risks associated with drilling by improving imaging of the subsurface, increasing oil recovery rates, and expanding the geographic scope of drilling; many were responses to the nationalization of production in the lowest-cost producing regions of the world in the Middle East, and enabled firms to increase production in the Gulf of Mexico, the North Sea, and elsewhere.

Another important explanation for economic liberalization focuses on the role of power in the international system. The United States, in particular, is argued to have used a mix of positive and negative incentives to open the economies of its trading partners. There is some historical evidence to support the role of the US government in promoting greater openness. Oil from the United States dominated global production in the years before World War II, and continues to make up a significant share of global supplies. The US-based majors also sought to use the diplomatic resources of the US government to help them expand their investment opportunities internationally in the 1920s and 1930s, including in the national markets of Japan and elsewhere.

Yet it is difficult to attribute recent changes in governance of the international oil market to pressure from the US government. Political support for opening markets in the service of US firms was complicated by the rise of Middle Eastern oil production, and the loss of competitiveness of US oil led to the imposition of import quotas. While these quotas were abolished in the 1970s, it remains unclear whether

[7] Jeffry Frieden and Ronald Rogowski, "The Impact of the International Economy on National Policies: An Analytical Overview," in *Internationalization and Domestic Politics*, Robert O. Keohane and Helen V. Milner, (eds.) (New York: Cambridge University Press, 1996), 25–47.

the preferences of US policymakers, or the domestic oil industry in the United States, favor abolishing barriers to trade and investment in the oil industry. Indeed, the US government continues to subsidize the production of oil domestically, and this policy is commonly justified under the rubric of promoting energy autarky. The private power of the US-based majors was also weakened in the 1970s with the rise of national oil companies, and the strategies of these firms began to focus on regaining access to oil production and cutting excess capacity rather than opening the national markets of other oil-importing states.

These are by no means the only explanations for economic liberalization that focus on forces external to domestic politics. In the chapters that follow I also address the role of externally driven crises, for example, in changing the strategies of firms and governments. Yet if we discount the role of falling transaction costs, and of power, in shaping changes in the governance of oil markets, then what other explanations might we also point to that can help to explain the changes that have occurred?

Domestic interests and oil market governance

Rather than focusing on transaction costs or power, others ascribe a more active role to domestic actors in explaining the changes made to economic governance over the past three decades. In the financial sector, for example, governments are argued to have promoted the globalization of financial markets.[8] Another possibility is that partisan differences explain patterns of liberalization. Parties of the right, for example, have been identified as favoring privatization, while those of the left are taken to favor government ownership of firms. Parties of the left are also argued to prefer increasing foreign direct investment because it facilitates inflows of patient capital.[9] It is also plausible

[8] Eric Helleiner, *States and the Reemergence of Global Finance: From Bretton Woods to the 1990s* (Ithaca: Cornell Univeristy Press, 1994); Rawi Abdelal, *Capital Rules: The Construction of Global Finance* (Cambridge, MA: Harvard University Press, 2007).

[9] Carles Boix, "Privatizing the Public Business Sector in the Eighties: Economic Performance, Partisan Responses and Divided Government," *British Journal of Political Science* 27 (1997): 473–96; Pablo M. Pinto and Santiago M. Pinto, "The Politics of Investment: Partisanship and the Sectoral Allocation of Foreign Direct Investment," *Economics & Politics* 20 (2008): 216–54.

that changes in the way political institutions aggregate interests have occurred in ways that reduced the influence of protectionist sentiment. In the United States, for example, shifting the control over trade policymaking to the president from Congress helped lower tariffs because the president has an incentive to reflect a broader constituency than the narrower interests represented by Congress. In the developing world, democratization is argued to have increased the voice of consumers relative to producers, thereby altering the policy equilibrium in favor of liberalization.[10]

The shift to more liberal forms of economic governance can be explained by the interests and strategies of firms, in addition to governments. The most recently cited case is finance, where firms are argued to have been crucial in forcing governments to lower barriers to capital flows internationally. The internationalization of production is also cited as leading industries to defend the liberalized trade regime against protectionist sentiment in the 1970s across a wide range of manufacturing sectors, even as firms' support for openness also appears to be contingent on the behavior of other countries because market imperfections can make success in the domestic market dependent on the ability to enter markets internationally.[11]

How might such explanations help us understand changes in oil markets? Many of these general approaches to analyzing political economic outcomes are echoed in studies of the politics of oil. Interest in oil among scholars, like automobile drivers, tends to follow the price of a gallon at the pump. There was a dearth of research on the politics of oil through the 1980s and 1990s when prices were low. In contrast,

[10] Susanne Lohmann and Sharyn O'Halloran, "Divided Government and US Trade Policy: Theory and Evidence," *International Organization* 48 (1994): 595–632; Fiona McGillivray, *Privileging Industry: The Comparative Politics of Trade and Industrial Policy* (Princeton University Press, 2004); Helen V. Milner and Keiko Kubota, "Why the Move to Free Trade? Democracy and Trade Policy in the Developing Countries," *International Organization* 59 (2005): 107–43.

[11] Simon Johnson and James Kwak, *13 Bankers: The Wall Street Takeover and the Next Financial Meltdown* (New York: Vintage, 2011); Helen V. Milner, *Resisting Protectionism: Global Industries and the Politics of International Trade* (Princeton University Press, 1988); Helen V. Milner and David B. Yoffie, "Between Free Trade and Protectionism: Strategic Trade Policy and a Theory of Corporate Trade Demands," *International Organization* 42 (1989): 239–72; Michael Luszig, "The Limits of Rent Seeking: Why Protectionists Become Free Traders," *Review of International Political Economy* 5 (1998): 38–63.

political scientists, economists, and historians examined the politics of oil in the wake of the price shocks of 1973–4 and 1978–9. Their claims and counterarguments are mirrored in the popular press, which veers between explaining oil policies as a function of "Big Oil," and proposing that oil is a strategic resource in which governments dominate policymaking.[12]

A number of studies of oil market governance paid particular attention to the interests of governments. There are good reasons for this: the rise of mechanized warfare in World War I made oil central to the ability of governments to prosecute war on the battlefield, and intervention in the energy sector is also often cloaked in the language of national security; there are numerous quotes from historical figures – from Winston Churchill arguing that "safety and certainty in oil lie in variety and variety alone" to Calvin Coolidge's claim that "the supremacy of nations may be determined by the possession of available petroleum and its products" – that suggest governments have independent interests in securing oil supplies because of the importance of oil products in military operations.[13]

Other studies show that the ability of the state to design and implement oil policies is constrained by domestic institutions. The response of US policymakers to oil price volatility of the 1970s, for example, was constrained by the weakness of US bureaucratic capacity relative to France and elsewhere, and this weakness forced policymakers in the United States to choose international diplomacy and price liberalization as their best response to the oil crises of the 1970s.[14]

[12] This statement refers to the politics of oil in the advanced industrialized states. Numerous studies have been published on the politics of producer country oil since the nationalizations of the 1970s. For a review see, Llewelyn Hughes and Phillip Lipscy, "The Politics of Energy," *Annual Review of Political Science* 16 (2013).

[13] Churchill is cited in Daniel Yergin, "Ensuring Energy Security," *Foreign Affairs* 85 (2006): 69–82. Coolidge is cited in Paul H. Frankel, *Essentials of Petroleum: A Key to Oil Economics*: (New York: Frank Cass, 1983). On the state-centric view see Stephen D. Krasner, *Defending the National Interest: Raw Materials Investments and US Foreign Policy* (Princeton University Press, 1978); Leslie E. Grayson, *National Oil Companies* (Chichester: John Wiley and Sons, 1981); Eric D.K. Melby, *Oil and the International System: The Case of France, 1918–1969* (New York: Arno Press, 1981). Also see Gerald D. Nash, *United States Oil Policy 1890–1964* (University of Pittsburgh Press, 1968).

[14] C. John Ikenberry, *Reasons of State: Oil Politics and the Capacities of American Government*, Cornell Studies in Political Economy (Ithaca: Cornell

A second problem with a state-centric view is that it ignores the fact that firms have interests in oil market governance, and these interests may not be aligned with governments. Recognizing this, a second family of explanations focuses on the interests of firms. Indeed, the seminal work on economic theories of regulation is motivated by the idea that oil policies in the United States implemented in the name of national security are a veil hiding the interests and strategies of firms. Firms, in turn, are assumed to support barriers to market entry in order to increase rents.[15] Studies on the politics of oil in the United States also focus on the interests of firms in oil market regulation. Even in countries such as France, which have historically been treated as having "strong states," the behavior of France's national oil champions is shown to have been shaped by private rather than public interests.[16]

Finally, a number of studies extend the interests of firms beyond national borders, arguing that national choices about how to regulate the domestic market are a function of the structure of international market governance, and pointing to the political influence of the oil majors, in particular, to explain outcomes. The best known of these studies is by Blair, who acted as advisor to the US Senate investigation into the international oil cartel that dominated oil politics for five decades.[17] French oil policy prior to and following World War I similarly reflected the battle for market share within France, which was an artifact of competition between the international majors over the control of oil production internationally. Instances of both *étatisme* and liberalization therefore emerge out of the strategies adopted by the majors, allied with domestic firms and financial interests, as they sought to put a floor on prices by limiting the production of crude globally.[18]

University Press, 1988); David F. Prindle, *Petroleum Politics and the Texas Railroad Commission*, (Austin, TX: University of Texas Press, 1981).

[15] George J. Stigler, "The Theory of Economic Regulation," *The Bell Journal of Economics and Management Science* 2 (1971): 3–21.

[16] Feigenbaum suggests, like Milner, that firms vary in their preferences towards economic liberalization. Harvey B. Feigenbaum, *The Politics of Public Enterprise: Oil and the French State* (Princeton University Press, 1985). Chubb and Oppenheimer also focus on industry interests in explaining oil market regulation in the United States. John E. Chubb, *Interest Groups and the Bureaucracy: The Politics of Energy* (Stanford University Press, 1983); Bruce Ian Oppenheimer, *Oil and the Congressional Process* (Lexington, MA: Lexington Books, 1974).

[17] John M. Blair, *The Control of Oil* (New York: Vintage, 1978).

[18] Gregory P. Nowell, *Mercantile States and the World Oil Cartel 1990–1939* (Ithaca: Cornell University Press, 1994). A more limited number of studies

Each of these arguments makes plausible cases about the interests of both governments and firms in oil market governance. Yet taken together they suffer from two shortcomings. First, while many incorporate evidence across national markets, few are truly comparative, and this lack of a comparative perspective can lead to an overstatement of the uniqueness of single-country attributes in shaping how oil markets are governed. While weak government institutions in the United States forced policymakers to choose price liberalization as the best response to oil market volatility in the 1970s, for example, oil market volatility led domestic actors to choose price liberalization in other countries, including France as I show in Chapter 5. In fact, firms and governments in many countries reordered the institutions governing oil markets in the 1980s and 1990s in response to structural changes in the oil market.

Single-country studies also lead scholars to treat important factors shaping differences in cross-national forms of oil market governance as constants, when they are in fact variables. Nowell's forensic analysis of the politics of oil in France, for example, does not explore how the nature of the settlements negotiated between the oil majors, domestic business, and government differed from country to country, as well as across time.[19] Indeed, it is this very difference in the division of spoils, I argue, that had crucial implications for the types of firms that emerged, which then went on to condition their responses to the transformation of the international oil market. While the comparative approach does

adopt a coalitional approach to analyzing the politics of oil market governance. Samuels, for example, argues that policy outcomes in Japan are a result of bargaining between policymakers and firms, in which firms seek autonomy from the state while seeking to increase economic rents through the use of regulation. Vietor focuses on the politics of oil market governance in the United States, finding that oil market regulation is a negotiated outcome between business and government. Cowhey argues that forms of governance are determined through a process of bargaining between firms and governments. Richard J. Samuels, *The Business of the Japanese State: Energy Markets in Comparative and Historical Perspective* (Ithaca: Cornell University Press, 1987); Richard H. K. Vietor, *Energy Policy in America since 1945: A Study of Business–Government Relations*, (Cambridge University Press, 1984); Peter F. Cowhey, *The Problems of Plenty: Energy Policy and International Politics* (Berkeley, CA: University of California Press, 1985).

19 Nowell recognizes, but does not delve into, differences in oil market governance across countries. See Nowell, *Mercantile States and the World Oil Cartel*, 220–1.

not allow us to address each case in the same level of detail, therefore, it does reveal the importance of heretofore unidentified factors that matter in explaining outcomes.

Stepping back from the benefits of a comparative approach, existing studies suffer from a third shortcoming: they do not explain how domestic actors responded to the transformative changes in the international oil market since the 1960s. This is not simply a matter of updating the historical record: the most important recent variation implied by this change is the weakening of the competitive position of the oil majors. Given that a goal of governments lay in limiting the corporate power of the majors, existing studies do not offer a sense of how firms and governments responded to this change. The nationalization of oil-producing assets would suggest a rise in intervention, yet we see the opposite. It is also not clear how firms responded to the new risks and opportunities associated with the development of a more liquid market for oil. One goal of this book is to answer these questions by examining how firms and governments responded to these changes in the international oil market, and in so doing reordered the structure of oil market governance.

Firms, industrial compacts, and oil market governance

In Chapter 2 I showed that the advanced industrialized states made changes to the laws and regulations governing national oil markets. I also found that governments in some of these countries continued to seek to protect and enhance the competitiveness of domestic firms, while others discarded this approach to governing oil markets. How can we explain these changes?

The first step in developing an explanation for these outcomes lies is recognizing that the oil sector is subject to the same kinds of distributional conflicts that occur in other sectors of the economy. This means we need to identify which actors are important in shaping oil policies, and account for their interests and the strategies they are likely to adopt. In this study I propose that choices about how to govern oil markets have both "demand" and "supply" elements, with outcomes the result of a process of bargaining between firms and governments. I pay particular attention to the interests and strategies of oil firms. While the problem of market power often places national security at the center of analyses of oil politics, this elides the fact that the most important actors investing and operating in oil markets are firms, and that policymakers have imperfect control over them. It also fails to recognize that the costs and benefits of intervening in oil markets are not limited to national security. Instead, the ubiquity of oil, and lack of perfect substitutes for many products derived from it, mean that oil market governance is tied to questions of competitiveness and growth. Below I outline firm interests and strategies, and then move on to discuss the interests of policymakers in oil market governance.

Firm interests and strategies

For firms, extracting favors from policymakers is not an end in itself. Rather, firms seek to maximize profits. Given this, the decision to lobby policymakers is determined by whether managers judge the distributive effects of government intervention to enhance profitability.[20] In the case of oil, firms often turned to the resources available to governments in seeking to improve their competitive position. A common problem historically, for example, lay in ensuring that new discoveries of oil did not lead to gluts that undermined prices. One corporate response in the United States and elsewhere was to integrate vertically. Vertical integration refers to "increasing the number of distribution and processing steps in an industry's cycle of activities that are under the ownership, management, or control of a single company."[21]

For the oil majors controlling supplies internationally and extending into national markets was judged the best strategy for stabilizing the supply of oil and for maximizing profits.[22] Integrating vertically ensured these firms could control the rate of output. Extending control in national markets then enabled these firms to dispose of oil

[20] For summaries of the view of policy as a resource for firms see Jean J. Boddewyn and Thomas L. Brewer, "International-business Political Behavior: New Theoretical Directions," *Academy of Management Review* 19 (1994): 119–43; Brian Shaffer, "Firm-level Responses to Government Regulation: Theoretical and Research Approaches," *Journal of Management* 21 (1995): 495–514; David Baron, "Integrated Strategy, Trade Policy, and Global Competition," *California Management Review* 39 (1997): 145–69.

[21] John G. McLean and Robert Haigh, *The Growth of Integrated Companies* (Norwood, MA: Plimpton, 1954), 8. For a general review see Martin K. Perry, "Vertical Integration: Determinants and Effects," *Handbook of Industrial Organization* 1 (1989): 183–255.

[22] Blair, *The Control of Oil*. See also J.H. Bamberg, *The History of the British Petroleum Company, Volume 2: The Anglo-Iranian Years, 1928–1954* (Cambridge University Press, 1994), 528–34.

produced; in the absence of external markets, internalizing production and refining functions increased the stability of supply and demand.[23] The strategy of vertical integration also provided its own justification by undermining the viability of other business models.[24]

The resources of government were useful in helping firms regulate supply. As I describe in Chapter 6, US firms commonly turned to the diplomatic resources available to the federal government, for example, to help them secure a share of oil production in the Middle East, the Dutch East Indies, and elsewhere. In the United States, which was the dominant producer globally until the 1950s, both federal and state regulations also played a central role in limiting the supply of oil domestically, and therefore onto the international market. The US Department of the Interior, for example, regulated interstate oil shipments in the United States, and made it an offense to ignore quotas set through the Texas Railroad Commission. The federal government also manipulated licenses to limit the amount of oil produced on federally owned lands.

Government was also useful to the non-majors in the struggle for market share within domestic markets, and in managing supply internationally. Faced with the dominance of the majors and no significant secondary market for oil and oil products, firms other than the majors, for example, had the choice of either tying up with a major, or seeking to protect their market share within the domestic market and secure access to the limited sources of crude oil available internationally that were outside their control. Government had the capacity to redistribute market share within the domestic market using quotas, licensing, and other policy instruments. Barriers to trade could increase the incentives to refine oil within the domestic market, while licensing procedures could be used to regulate the rights to build new capacity. Governments also had the ability to subsidize investments both domestically and internationally through tax measures, loan guarantees, and direct subsidies. Considering this, it is unsurprising that firms actively sought to use the resources of government.

23 On Standard Oil (New Jersey) integrating forwards into production see Alfred D. Chandler, Jr. *Strategy and Structure: Chapters in the History of the American Industrial Enterprise* (Cambridge, MA: MIT Press, 1969[1962]), 163–224.

24 Nick Antill and Robert Arnott, *Oil Company Crisis: Managing Structure, Profitability, and Growth* (Oxford Institute of Energy, 2002), 5–6. Paul Stevens (ed.), *Strategic Positioning in the Oil Industry: Trends and Options* (Abu Dhabi: The Emirates Center for Strategic Studies and Research, 1988).

The transformation of the international oil market described in Chapter 2 changed the balance of risks and opportunities facing firms. In the short run, expropriations undermined the profitability of many firms, although this was mitigated by the increase in crude oil prices. High prices also harmed profits by undermining the demand for oil products other than in the transportation sector, where there are few substitutes for gasoline and diesel. It also changed the structure of demand for different oil products.

In the long run, however, the loosening of the majors' control over international trade in oil had two implications for firm strategy, and by extension for the demands they made to governments. First, the fragmentation of the oil market diversified the range of opportunities available to firms seeking to expand into new markets.[25] In particular, the loss of control by the majors, coupled with the relaxation of trade barriers in the US following the peaking of US production, further facilitated access to the crude production stage. This promised greater profits because the geological fact of the limited dispersion of oil internationally implies that, over the long run, profits will be higher in the production of oil than in refining, where the barriers to entry are lower. Shifting greater resources into the production phase therefore meant increasing profits. It also reduced the importance of operations in any single national market, given that firms had more opportunities to buy and sell crude and crude products internationally.

Second, the change in oil market structure increased price volatility. Rather than being determined by the majors unilaterally, oil shifted over time to a system of formula-based pricing, in which the spot-market is used to price long-term and short-term contracts, with the price discounted to take account for differences in the quality of crudes.[26] For firms, this volatility increased the incentives to diversify operations both geographically and across different business segments. Selling refined products across multiple markets, for example, reduced the reliance on underlying growth in any single market. In exploration and production, operating across multiple jurisdictions and projects

[25] The abolition of quotas on imports in the United States also affected the range of opportunities available to firms.

[26] OPEC administered prices until the mid-1980s using a variety of strategies, before the market standard shifted to formula pricing, beginning with PEMEX in 1986. Bassam Fattouh, "An Anatomy of the Crude Oil Pricing System, *Oxford Institute for Energy Studies WPM* 40 (January 2011)."

spread political risk, as well as the technological risks associated with the complexity of each individual project.[27]

While the transformation of the oil market changed the incentives facing firms, their ability to exploit this change depended on firm characteristics, and I argue this had important implications for the demands they placed on governments. Firms that enjoyed a degree of scale and vertical-integration were in a better position to take advantage of the shift in the structure of the international oil market by diversifying across business segments and geographic markets. Diversification offered an opportunity to reduce the dependence on a single national market, meaning losses in any given market could be balanced against profits made in others. Indeed, one market-based response adopted to the transformation in the oil market was to reduce the degree of operational integration within firms, enabling business units operating within single business segments within the firm to sell crude oil and refined products externally, as well as in different markets.[28] Diversification also enabled these firms to better manage the problem of price volatility. Even for the majors that lost control over significant productive assets through the nationalizations of the 1970s, diversification was the key to managing these new risks and opportunities.

I argue this had important implications for the utility of government intervention. Given the changes in oil market structure, as these firms further diversified they became less interested in investing resources in maintaining barriers to trade.[29] The usefulness of financial resources provided by the government also fell as firms grew in size and government budgets became increasingly constrained. In terms of lobbying behavior, this meant these firms invested fewer resources in maintaining the status quo, and became more willing to acquiesce to reform proposals by others. When existing laws and regulations undermined

[27] Antill and Arnott note a range of other benefits to remaining organized across business segments, including lower costs of capital and tax minimization. Nick Antill and Robert Arnott, "Creating Value in the Oil Industry," *Institutional and Corporate Finance* 16 (2005): 18–31. The benefits of risk diversification were identified as conducive to vertical integration much earlier. McLean and Haigh, *The Growth of Integrated Companies*, especially 115.

[28] Paul Stevens, "Economists and the Oil Industry: Fact versus Analysis, the Case of Vertical Integration," in *Energy in a Competitive Market: Essays in Honour of Colin Robinson*, ed. Lester C. Hunt (Cheltenham: Edward Elgar, 2003), 95–102; Antill and Arnott, "Creating Value in the Oil Industry."

[29] Milner, *Resisting Protectionism*.

their ability to profit from the new opportunities, and manage the risks, this made these firms actively pursue reform.

The shift in firm strategies also had an important effect on outcomes. The choices facing firms as they develop strategies towards governments are often portrayed as binary, either actively supporting or opposing protection. Yet a third possibility exists: that firms simply cease investing resources in retaining government support. Given that laws and regulations have important distributional effects, benefitting some firms while imposing costs on others, the increasing unwillingness of more diversified and larger firms to invest in maintaining the status quo meant actors that oppose the status quo were more likely to be successful.[30] In the empirical chapters I also show that oil firms actively pursued liberalization when existing laws and regulations undermined profits sufficiently to overcome the costs of lobbying.

Conversely, firms that remained focused on single segments were less well positioned to take advantage of the opportunities presented by the transformation of the international oil market. They were also less able to manage the risks associated with oil price volatility. What does this mean for these firms' approach towards government? Most simply, for firms focused in single segments the resources available from government continued to be valuable, given they remained less diversified and therefore less able to manage volatility across different markets and products. They were also more willing to seek protection within the domestic market. Smaller firms were also more likely from seek the support of governments in the form of capital as they sought to diversify operations internationally and across business segments, in order to improve profitability.[31] While they were unlikely to win all of their demands, their continued demand for government support increased the likelihood this would remain a component of oil market governance.

The primary focus of this study is on the types of demands oil firms placed on policymakers. A complete explanation nevertheless

30 Jack Knight, "Models, Interpretations, and Theories: Constructing Explanations of Institutional Emergence and Change" in *Explaining Social Institutions*, eds. Jack Knight and Itai Sened (Ann Arbor, MI: University of Michigan Press, 1995), 95–120.

31 Antill and Arnott note the relationship between firm structure and the ability to raise financing. Antill and Arnott, "Creating Value in the Oil Industry," 24.

requires us to account for changes in governments' incentives to supply support to firms. Here an important consideration is a second characteristic of oil: products derived from it are used as an input in transportation and manufacturing, and are also a pocketbook issue for voters.[32] If we make the simple assumption that policymakers care about their chances for reelection, and economic performance matters to their chances of being reelected, then if follows that the strategies they choose to shape oil markets cannot ignore the implications of oil prices for other sectors of the economy. Indeed there is evidence that oil prices can have an important effect on macroeconomic performance, although this effect has fallen since the 1970s as a result of the substitution of oil by natural gas, nuclear power, and other energy sources in electricity generation.[33]

In addition to the broad economic effects of higher oil product prices, industries with low price elasticity of demand for oil products, and consumers, are also likely to oppose policies that increase the costs of crude products. Rather than a monotonic function of the demands of oil firms, or policymaker concerns about market power or national security, therefore, governments face a trade-off in supplying support to firms as they seek to increase the welfare of the population, and competing claims from different industries.[34]

The origins of firm characteristics

While firm incentives are influenced by changes in the structure of markets, this does not tell us why some firms succeed while others do

[32] Kindleberger notes that the politics of intermediate goods such as oil products differ from those of final goods. Charles P. Kindleberger, "The Rise of Free Trade in Western Europe, 1820–1875," *The Journal of Economic History* 35 (1975): 20–55. Ray suggests that consumers should be more sensitive to policies that directly affect their well-being. Edward John Ray, "The Determinants of Tariff and Nontariff Trade Restrictions in the United States," *Journal of Political Economy* 89 (1981): 108.

[33] James D. Hamilton, "Oil and the Macroeconomy since World War II," *The Journal of Political Economy* 91 (1983): 228–48; J. Peter Ferderer, "Oil Price Volatility and the Macroeconomy," *Journal of Macroeconomics* 18 (1997): 1–26; Olivier J. Blanchard and Jordi Galí, "The Macroeconomic Effects of Oil Price Shocks: Why are the 2000s so Different from the 1970s?" *NBER Working Paper No. W13368* (September 2007).

[34] Kishgore Gawande, Pravin Krishna, and Marcelo Olarreaga, "Lobbying Competition over Trade Policy," *NBER Working Paper 11371* (2005).

not. A common method for accounting for differences in the competitiveness of industries and firms cross-nationally, and by extension the strategies they adopt towards government, points to countries' relative factor endowments. Under the specific-factors model, for example, the opening of markets is proposed to benefit industries that use abundant factors intensively. If we extend this to firm strategy, this implies these firms should support market-opening measures. Industries that use scarce factors intensively, on the other hand, are expected to suffer from the introduction of market-opening measures. By extension, this should lead them to support policies of protection, and oppose liberalizing measures.[35]

In the case of oil, however, it is difficult to use this to account for the competitive position of firms. The advanced industrialized countries, with the exception of a handful of countries, are poorly endowed with reserves of oil. Data on global oil production collected by BP shows, for example, that of the OECD members only Denmark, Norway, the United States, Canada, Australia, and the United Kingdom produced a daily mean in excess of 100,000 barrels between 1980 and 2005.[36] Assuming similar levels of physical and human capital across these countries, this makes factor endowments an implausible explanation for relative levels of competitiveness across firms, and by extension their position towards economic liberalization. Yet, despite this, we see a significant number of diversified firms operating at scale in these countries, even while some firms remain small and largely focused in single market segments, often within the domestic market.

Political economists have long noted that differences in national institutions can lead to different choices over the instruments used to support firms.[37] While managerial prowess undoubtedly affects cross-national variation in firm characteristics, I argue that in the case of oil, a series of industrial compacts negotiated between business interests and governments in earlier periods crucially shaped

[35] Alt et al., "Asset Specificity and the Political Behavior of Firms," 99–116.

[36] Data drawn from BP Statistical Review of World Energy 2010.

[37] Andrew Shonfield, *Modern Capitalism: The Changing Balance of Public and Private Power* (New York: Oxford University Press, 1965); Peter J. Katzenstein, "Introduction: Domestic and International Forces and Strategies of Foreign Economic Policy," *International Organization* 31 (1977): 587–606; John Zysman, *Government, Markets, and Growth: Financial Systems and the Politics of Industrial Change* (Ithaca: Cornell University Press, 1983);

the characteristics of firms, and therefore the types of policies they went on to demand from policymakers. These compacts shared two characteristics. First, they regulated market share between firms through the application of policy instruments, including both tariff and non-tariff measures. They often included efforts to regulate market shares by manipulating the size and competitive position of firms, and were commonly formalized in laws that included sanctions for non-compliance. Second, they were durable, with many established in the interwar years and lasting, with only incremental changes, until the 1980s.

Rather than reflecting the interests of firms alone, or the interests of government, these industrial compacts emerged through a process of bargaining between business and governments, in which each had an independent set of interests. The strategic nature of this bargaining meant these compacts also tended to differ across countries and across time. In the early process of bargaining over the division of market share in France, for example, which I take up in the next chapter, policymakers had few tools through which to respond to the encroachment of the majors into the domestic market. There were also divisions between the left and the right over the best response. This absence of a well-developed bureaucratic structure responsible for oil policy planning did not mean policymakers had no effect on the nature of the institutions that emerged, but it did put them in a weak position, relative to the majors and other business interests, in designing and implementing policy. Certainly their position was weaker than in the postwar period, when oil policymaking was absorbed into a

Peter A. Hall, *Governing the Economy: The Politics of State Intervention in Britain and France* (New York: Oxford University Press, 1986); Frank Dobbin, *Forging Industrial Policy: The United States, Britain, and France in the Railway Age* (Cambridge University Press, 1994); G. John Ikenberry, "History's Heavy Hand: Institutions and the Politics of the State," Paper prepared for a conference on "New Perspectives on Institutions," University of Maryland, October 1994. Also see Michael E. Porter, *The Competitive Advantage of Nations* (New York: Free Press, 1998 [1990]). More recently, the varieties of capitalism approach to political economy takes a firm-centered approach to understanding differences between political economies, analyzing how institutions shape the way in which firms contract with others. Peter A. Hall and David Soskice, "An Introduction to the Varieties of Capitalism," in *Varieties of Capitalism: The Institutional Foundations of Comparative Advantage*, eds. Peter A. Hall and David Soskice (Oxford University Press, 2001), 1–68.

broader bureaucratic structure that aimed to promote French industrial growth.[38]

In the Japanese case, which I take up in Chapter 5, the Japanese government was also forced to respond to the encroachment of the majors into the domestic market. The industrial compact negotiated in Japan differed from that of France, however, because of how domestic firms responded to the idea of a state monopoly in oil, and because of the different degrees of influence of the central government and military within Japan proper and its colonies. In the postwar period, the institutions governing oil markets were renegotiated as part of a broader project of promoting industrial growth, like in France, however, firms once again largely resisted attempts to force mergers in order to create a larger and more vertically integrated firm capable of competing internationally.

Rather than emerging as part of a planning process – with firms acting as agents for their principals in government – or reflecting the interests of firms alone, these compacts therefore emerged from a process of political bargaining, in which both firms and governments sought to shape the institutions governing oil markets in their favor. The differences in them were crucial, however, in shaping the characteristics of the firms operating in the oil sector in each of the countries examined in this study, and therefore the strategies each adopted in response to changes in the structure of the international oil market.[39]

Alternative explanations: the role of ideas and the influence of Europe

There are two families of theories that offer plausible alternatives to the account offered here, even if we discount falling transaction costs

[38] On bureaucratic structure see Paul Pierson, "Review: When Effect Becomes Cause: Policy Feedback and Political Change," *World Politics* 45 (1993): 604. Ikenberry argues that the lack of a bureaucratic elite in the United States hampered the ability of government to respond effectively to the oil crises of the 1970s. G. John Ikenberry, "The Irony of State Strength: Comparative Responses to the Oil Shocks in the 1970s," *International Organization* 40 (1986): 105–37.

[39] Knight, "Models, Interpretations, and Theories"; Jack Knight, "Explaining Neoliberalism," in *The Rise of Neoliberalism and Institutional Analysis*, eds. John L. Campbell and Ove K. Pedersen (Princeton University Press, 2001), 38. Also see Terry M. Moe, "Political Institutions: The Neglected Side of the Story," *Journal of Law, Economics, & Organization* 6 (1990): 213–53.

and the role of power in explaining changes in oil markets. Conducting an exhaustive analysis of each is impossible within the confines of a single book. I nevertheless address them explicitly in the chapters that follow, given their plausibility as alternative accounts.

The first is rooted in the role of ideas in shaping the strategies chosen by policymakers to govern economic markets. Scholars point out that, in the wake of the economic crisis of the late 1970s, policymakers shifted their beliefs about the best way to govern economies. Not only in energy markets, but across manufacturing sectors, governments withdrew from actively shaping industrial outcomes in the name of economic growth and full employment. This change was informed by the failure of the previous economic model to provide tools to manage the economic crisis facing governments.[40] Stevens has argued this is also the case in the oil industry, where changed understanding of appropriate forms of government intervention in economic markets affected outcomes in the oil industry.[41]

How might ideational change affect forms of governance in oil markets? In addition to the general change in beliefs about appropriate forms of economic governance, a plausible source of ideational change lies in a shift in the way policymakers understood the nature of risk in the international oil market. The initial view of the oil market identified petroleum's exhaustibility as an important long-run driver of supply. The non-renewable nature of natural resources was first elaborated in Harold Hotelling's 1931 study.[42] Hotelling proposed that resource prices, including in the case of petroleum, are determined

[40] Mark Blyth, *Great Transformations: Economic Ideas and Institutional Change in the Twentieth Century* (Cambridge University Press, 2002); Peter A. Hall, "Policy Paradigms, Social Learning, and the State: The Case of Economic Policymaking in Britain," *Comparative Politics* 25 (1993): 275–96; Scholars also propose ideational change is influenced by lessons learned in similar countries. Beth A. Simmons and Zachary Elkins, "The Globalization of Liberalization: Policy Diffusion in the International Political Economy," *American Political Science Review* 98 (2004), 171–89.

[41] Stevens, *Strategic Positioning in the Oil Industry*.

[42] Harold Hotelling, "The Economics of Exhaustible Resources," *The Journal of Political Economy* 39 (1931): 137–75. For an early discussion of the economic characteristics of exhaustibility of resources see Lewis Cecil Gray, "Rent under the Assumption of Exhaustibility," *The Quarterly Journal of Economics* 28 (1914): 466–89. Devarajan and Fisher summarize the development of Hotelling's theory. Shantayanan Devarajan and Anthony C. Fisher, "Hotelling's 'Economics of Exhaustible Resources': Fifty Years Later," *Journal of Economic Literature* 19 (1981): 65–73. Solow offers a summary

by the problem of exhaustibility. This view was adapted by M. King Hubbert, who predicted the peaking of US oil production in the lower-48 states. According to his theory, supplies should peak as oil becomes increasingly scarce, before going into irrevocable decline.[43]

The first tests of Hubbert's framework applied to world oil markets came with the wave of empirical research following the oil crises of the 1970s. It found little evidence of an upward price trend for crude oil prior to 1970, and proposed an alternative model of oil market dynamics in response. Here exhaustibility is a "geological fact of no economic interest." Instead, supply is determined by levels of investment, and an important variable affecting price is producer market power.[44] This remains the dominant paradigm for understanding the oil market today, as shown by its prevalence in microeconomic textbooks.[45]

The shift in understanding about the dynamics of supply in oil markets was underpinned by the material changes in the oil market outlined in Chapter 2. While policies designed to enhance the commercial position of domestic firms may have had some utility up to the 1960s, the increase in the distribution of oil production globally, and creation of external markets for oil and oil products, means there is now little

of the theory of exhaustible resources in his 1973 Ely Lecture: Robert M. Solow, "The Economics of Resources or the Resources of Economics," *The American Economic Review* 64 (1974): 1–14.

43 M.K. Hubbert, "Energy Resources: A Report to the Committee on Natural Resources," National Academy of Sciences, Government Printing Office, Publication No. 1000D (1962). On the difference between Hotelling and Hubbert see G.C. Watkins, "Oil Scarcity: What Have the Past Three Decades Revealed?" *Energy Policy* 34 (2006): 508–14.

44 Morris A. Adelman, "Mineral Depletion, with Special Reference to Petroleum," *The Review of Economics and Statistics* 72 (1990): 1–10. See also Morris A. Adelman, "Scarcity and World Oil Prices," *The Review of Economics and Statistics* 68 (1986): 387–97. This does not mean there was consensus among economists about the causes of the rapid increase in petroleum prices in the 1970s. A number of scholars contended, for example, that price shocks followed a path best explained by a rise in petroleum demand and short-term losses of supply, coupled with the short-term inelasticity of demand, rather than any exercise of market power by the OPEC cartel. See Dermot Gately, "A Ten-Year Retrospective: OPEC and the World Oil Market, *Journal of Economic Literature* 22 (1984): 1100–14."

45 For a standard economics textbook approach see N. Gregory Mankiw, *Principles of Economics*, 3rd edn. (Mason, OH: Thomson/South-Western, 2004), 353.

risk that contracted supplies of oil will not be honored. Given this, it is plausible that policymakers came to understand that liberalization, coupled with strategic oil reserves as a physical hedge, represented an alternative and more efficient method for managing any national security risks associated with the supply of oil.

The ideational hypothesis leads to a different set of expectations to the explanation above in terms of both who matters and what they want. Most obviously, it proposes that policymakers are central to choices about how to govern oil markets. The strategies of firms, in this argument, are secondary to policymakers and their preferred method for governing oil markets. Showing that policymakers matter more than firms is necessary, but not sufficient, to demonstrate that the ideational hypothesis best captures the cause of changes, however. If this argument is right, we should also find evidence about how these policymakers chose their preferred strategies for governing oil markets. In particular, there should be evidence that policymakers shifted their preferred strategy as a result of the input of new information about the nature of risk associated with relying on international markets for the supply of oil. Here we should look for evidence from policymakers, or the organizations that have delegated responsibility for oil policymaking, that they have absorbed these lessons about the changing structure of risk in the international oil market and sought to implement these in policy.

There is a second plausible alternative explanation for variation in institutions of governance in oil markets. Europe has become increasingly involved as an independent actor in member states' policies in energy markets, including in the oil sector.[46] Further, a large number of the major oil-importing states within the OECD are located in Europe. In the 1950s and 1960s the European institutions failed to significantly influence the national energy policies of member states, with market integration taken as a secondary priority behind managing externalities, including ensuring security of supplies. Yet since the late 1960s the European Commission has consistently argued for reducing barriers to trade and freeing investment rules between European states in order to complete the single market within Europe. This pressure

[46] Stephen Padgett, "The Single European Energy Market: The Politics of Realization," *Journal of Common Market Studies* 30 (1992): 53–76.

increased with the agreement of the Council in 1987 to identify the barriers to completing the internal market in energy products.

The influence of Europe therefore represents an alternative explanation for outcomes in European states, given that all but two of the major oil-importing states are located in Europe, even if it cannot explain outcomes across all the major oil-importing states. What should we expect if the influence of European institutions matters in explaining the broad changes outlined in the previous chapter? Most simply, we should see that European organizations promote liberalization and privatization in oil markets. We should also see some evidence that the policy positions taken within Europe have an effect on national choices about oil market governance. By extension, this suggests that the interests and strategies of firms and policymakers should be relatively unimportant, and that changes should be implemented even when they are in opposition to national policymaker and firm interests.

Empirical strategy

The goal of this book is to explain changes in the way the advanced industrialized states governed oil markets. In order to assess these claims I conduct a historical and comparative institutional analysis of oil market governance in France, Japan, and the United States.[47] Differences in political institutions and other sources of variation in national economies mean it is challenging to isolate the effects of industrial compacts on firm characteristics, and the relative importance of firms in influencing forms of oil market governance within a cross-national setting. The choice of France and Japan as comparative cases is nevertheless appropriate because it shows variation in the variables of interest while holding a number of other characteristics constant. Most importantly, both France and Japan have negligible supplies of oil.[48] We can safely assume, therefore, that any variation

[47] Greif notes that historical and comparative institutional analysis is a useful method for investigating how "past institutional, economic, political, social, and cultural features interact in shaping the nature of contemporary institutions and their evolution." Avner Greif, "Historical and Comparative Institutional Analysis," *The American Economic Review* 88 (1998): 80–4.

[48] France had colonial possessions in North Africa that proved to be rich in oil, but this difference with Japan should not be overstated. These reserves

should not be able to be explained as a function of differences in initial resource endowments. Further, although differences in the demand for oil products could plausibly affect firm characteristics, both countries saw a rapid increase in demand in both the industrial and transport sectors along with industrialization. While France was more successful than Japan in introducing nuclear energy, which is a partial substitute for oil in electricity generation, both have substantial nuclear energy programs and continue to rely almost wholly on oil products in the transport sector. Finally, the relative closeness of relations between firms and government historically also allows us to draw a direct line between the interests and strategies of these groups, and policy outcomes.

In terms of explanatory variables, on the other hand, French firms Total and Elf Aquitaine became large and diversified firms prior to the 1980s, while Japanese firms operating in the oil sector are historically small and weakly diversified. Further, as I show in the following chapters, this variation can be attributed to differences in the industrial compacts negotiated between business and governments in earlier periods.

The case of the United States, which I take up in Chapter 6, usefully extends the study of firms and oil market governance in France and Japan in two ways. First, although industry–government links are tight in both France in Japan, making it plausible to draw a line from their interests to policy outcomes, it remains possible that differences in political institutions changed the way the interests of these actors are weighted in shaping outcomes, and could also plausibly affect the strategies that firms adopt. The case of the United States, on the other hand, allows us to examine how differences in firm characteristics affect the strategies they adopt towards government in a setting in which national political institutions do not vary. This is because the United States offers significant variation in firm characteristics, with major international oil firms – in the form of Exxon and Chevron – that are large and diversified, operating alongside small- and medium-sized firms that have no international presence and focus on single segments of petroleum production, as well as refiners focused on producing products for the domestic market. Second, the case of the United States

were developed late, at a time when Japanese firms also obtained access to substantial oil supplies internationally, and were soon nationalized.

allows us to examine the interaction between initial resource endowments and industrial compacts on firm characteristics, although this introduces some complexity into the range of instruments available to governments to promote firm interests, as well as in interest group politics, that is not evident in the French and Japanese cases. I take up these issues in greater detail in Chapter 6.

Challenges

It is worth noting two challenges with the empirical strategy adopted here. First, while a comparative historical analysis is well suited to examining the effects of institutions on firm characteristics, and tracing how the strategies of firms influenced outcomes, it is less well suited to weighing the relative importance of firm strategies and other factors in shaping outcomes across the entire universe of cases. I return to discuss this later, but it is worth noting that the cases selected here do give us some confidence that they can be extended to other cases. While the international oil market is global, for example, regional differences in price suggest that the oil market may not be "one big pool" as it has been characterized.[49] The cases examined here are drawn from the three major oil-consuming regions of the world, however, ensuring that no bias is introduced because of characteristics associated with regional oil markets that alter the risk profile facing government, and therefore lead them to pursue different forms of governance. Second, this case selection ensures that the results are not skewed by European states. The mean change in patterns of oil market governance is away from intervention and towards more liberal forms of governance. It is also the case that European states make up the majority of the advanced industrialized states, as noted above. In addition to explicitly taking up the plausibility of a Europe-centered explanation in Chapter 4, selecting the United States and Japan as cases for examination offers

[49] This question has been subject to some debate. See Robert J. Weiner, "Is the World Oil Market 'One Great Pool'?" *The Energy Journal* 12 (1991), 95–108; Morris A. Adelman, "Is the World Oil Market 'One Great Pool'? – Comment," *The Energy Journal* 13 (1992): 157–8; A.E. Rodriguez, and Mark D. Williams, "Is the World Oil Market 'One Great Pool'? A Test," *Energy Studies Journal* 5 (1993): 121–30.

a second way of examining the validity of the argument advanced in this chapter.[50]

A second challenge lies in the problem of simultaneity. If we conceive of changes in oil market governance as a form of institutional change, an important problem lies in disentangling the effect of institutions on the strategies of firms from the effects of firm strategies on institutions. The empirical strategy here does not offer a perfect method for addressing this problem. Nevertheless, the transformation of the structure of the international oil market suggests that we can understand the primary direction of influence in the period prior to the 1980s being from industrial compacts to firm characteristics, while in the second period, when firms and government renegotiated forms of intervention, the most important direction of influence points from firm strategy to the renegotiation of these industrial compacts. Thus we can treat these as two distinct periods in analyzing the relationship between industrial compacts, firm characteristics, and forms of oil market governance.

Measuring oil market governance, and firm and policymaker strategies

This book is focused on explaining changes in oil market governance over the past three decades. Given this, appropriate indicators of changes in oil market governance are changes made to the laws, regulations, and policies implemented in oil-importing states, which shape market behavior. In Chapter 2 I focused on changes in the application of trade instruments, state ownership of national oil companies, and – in the case of the United States – laws and regulations designed to increase the share of domestic production. These remain the focus of the empirical chapters that follow, although I also pay attention to a broader set of indicators, which becomes possible given the greater detail with which policies are able to be examined in case study research. Extending consideration to regulatory and fiscal

[50] Searight and Gerring refer to this as the "diverse" case selection strategy, making the cases representative in the sense of representing variation within the population. The drawback is that it does not necessarily reflect the overall distribution across the population. See Jason Seawright and John Gerring, "Case Selection Techniques in Case Study Research: A Menu of Qualitative and Quantitative Options," *Political Research Quarterly* 61 (2008): 297.

instruments also enables us to examine whether the reduction in the use of trade instruments, and in the privatization of national oil firms, occurred because of a substitution effect, in which the use of these policy instruments were replaced in favor of substitutes such as tax or other regulatory policies.

I use secondary literature to identify the influence of policymakers on outcomes prior to the transformation of the international oil market. There are a number of excellent studies of the early period of oil market governance that utilize primary source material for each of the cases. These sources, in addition to primary source materials employed, are adequate to substantiate the general claim that the industrial compacts negotiated in earlier periods shaped firm characteristics.

For firm characteristics, I focus on the degree of firm diversification and size. Diversification is understood both geographically – in terms of internationalization – as well as by the number of market segments operated in. Data are drawn from annual reports, internal and public documents, and national and European assessments of the state of the industry, where available. I also use these sources to understand firms' market strategies. In order to identify firm strategies towards government, I use hearings in which industry associations make statements about their preferred policy outcomes. Firms typically do not publicly detail lobbying efforts. Industry associations, on the other hand, are designed to act in the interests of member companies. As such they have an explicit role in the policy formulation process. Further, given their role as representative of firm interests in the political process, using industry association policy preferences as data does not significantly undermine the validity of their statements as indicative of underlying firm preference. Where available I also utilize firm-level information. Senior executives often refer to proposed policy changes in letters to shareholders contained within annual reports, for example. They also make statements at congressional and parliamentary hearings, and I make use of this information when it is available. In the case of France, direct evidence of the policy preferences of firms is used given the dominance of the Total and Elf Aquitaine groups within the French petroleum sector. I use these sources to identify firm strategies, as well as statements and positions announced by major industry associations. In the French case, data used to identify European policy preferences are drawn from primary sources obtained from the European Commission and Council of Ministers, in addition to interviews with firm and state actors, and secondary sources. Finally, this data is complemented by open-ended interviews with firm representatives, representatives from the relevant political appointees, and representatives from bureaucratic agencies in charge of decision-making in Japan, France, and the United States, with the goal of better understanding the process that led to changes towards oil market governance since the 1980s.

4 *Transforming French oil market governance*

We were like priests not believing in God anymore.

Gilles Bellec
Director of Hydrocarbons (1984–9)
Ministry of Industry

This chapter documents changes made to oil market governance in France. France entered the 1980s with laws that regulated domestic market share, and with the government holding equity in two domestic oil firms. Yet by the end of the period examined in this study the structure of oil market governance was transformed. The national law regulating trade in oil and oil products – which had been part of the governance of the French oil market since 1928, except for during the interlude of war – was replaced by the free movement of oil and oil products across national borders. Prices on oil products were liberalized, and the two oil firms in which the government retained shareholdings and managerial representation – Elf Aquitaine and Total – were privatized, and merged in 2000 to create the fourth largest global integrated oil major by market capitalization.

The robustness of this transformation was demonstrated by a new French energy law passed in 2005. French policymakers, like those in Japan, responded to the rise in long-run oil prices, and the growing importance of climate change, by reconfiguring the national energy strategy. New measures were designed and implemented in order to shape the supply and demand for energy, but there was no demand for a return to policies favoring national champions. This suggested that the longstanding goal of regulating the market share of national firms was dead. In this chapter I explain how this change came about.

I proposed in the previous chapter that an important cause of economic liberalization in oil markets was a shift in firm strategy. In this chapter I show that the commercial importance of the domestic market fell over time relative to other markets for France's oil companies. This meant the firms were less interested in investing resources in retaining the laws and regulations that had defined relations between business and government in the oil sector since the interwar years. The shift in firm interests reflected the greater opportunities presented in the international oil market, allowing them to diversify their sources of production and enter downstream markets in North America and elsewhere. Indeed, government ownership emerged as a constraint as the firms grew and diversified, and French government became increasingly indebted.[1] When questions of liberalization and privatization were raised, therefore, firms were either indifferent to the proposed changes, or actively lobbied to shift the institutional equilibrium when it no longer met their needs.

Institutional changes are rarely caused by single events. As I describe in this chapter, liberalization of the French oil market was influenced by policymakers' response to slower growth, inflation, and rising debt, which led to a greater acceptance of markets for allocating goods and services.[2] There was also substantial pressure from new entrants into the oil sector to liberalize prices and the regulations governing imports and exports. Yet the embrace of liberalization is not sufficient to explain outcomes in the oil sector. A contrast with the French electricity and gas sectors is instructive. The energy sector has always been treated apart from other sectors of the economy. European efforts to complete the terms of the Treaty of Rome by opening borders to trade in energy products was unsuccessful for decades, and while European recommendations to liberalize the gas and electricity markets were made concurrently with those in oil, the former proceeded far more slowly.

1 Michael Loriaux, *France after Hegemony: International Change and Financial Reform* (Ithaca: Cornell University Press, 1991). I thank Jonah Levy for the second of these points.

2 Serge Bellanger and Olivier Blanchard, "Economic Policy," in *Transformations in French Business: Political, Economic, and Cultural Changes from 1981–1987*, eds. Judith Frommer and Janice McCormick, (Westport, CT: Greenwood Press, 1989), 35–50; Vivien A. Schmidt, *From State to Market? The Transformation of French Business and Government* (Cambridge University Press, 1996); Jonah D. Levy, "Economic Policy and Policy-Making," in *Developments in French Politics 3*, eds. Alistair Cole, Patrick Le Gales, and Jonah D. Levy (Basingstoke: Palgrave Macmillan, 2005), 170–95; Pepper D. Culpepper, "Capitalism, Coordination, and Economic Exchange: The French Political Economy since 1985" in *Changing France: The Politics that Markets Make*, eds. Pepper D. Culpepper, Peter A. Hall, and Bruno Palier (London, UK: Palgrave Macmillan, 2006), 29–49.

The French state retained majority ownership of the state electricity utility Électricité de France (EDF) after the reform of its relationship with the state in 2004, as well as in the national champion in the gas sector, Gaz de France, until 2008 when it merged with Suez. The government also retained public ownership in the nuclear energy sector through its seventy-three percent stake of nuclear supplier AREVA through the Commissariat à l'énergie atomique et aux énergies alternatives, as well a further ten percent held directly. Given this, we need to take into account the change in the strategies of firms operating in the oil sector, as well as their supporters within government, in order to explain outcomes.

In the first part of this chapter I describe how the industrial compact negotiated between business and government shaped the characteristics of French firms operating in the oil sector. Given the increasing importance of the European Commission, Directorate General of Competition, and European Court of Justice in influencing French policymaking, and the significance of Europe as an alternative explanation for outcomes, I also describe the preferred policies sought by European actors in the petroleum sector.

Oil market governance in France

The politics of oil in France is often framed by the national security interests of the state. Yet the usefulness of government instruments in regulating competition between the oil majors and a coterie of indigenous firms was at least as important in setting the terms of oil market governance after World War I.[3] Prior to the 1920s there was little coordinated national policy governing oil, and trade in oil and oil products were governed by the market. Imports were initially overwhelmingly sourced from the United States and Russia. French customs registered the first import of oil and products from the United States in 1861 and from Russia in 1866, although Russian oil only

[3] For a detailed analysis of the origins of the French domestic refining industry see Gregory P. Nowell, *Realpolitik vs. Transnational Rent-Seeking: French Mercantilism and the Development of the World Oil Cartel, 1860–1939* (Cambridge, MA: Massachusetts Institute of Technology, 1988); Daniel Murat, *L'intervention de l'État dans le secteur pétrolier en France* (Paris: Éditions Technip, 1969), 7–8.

became important from 1893.[4] Customs and tariff acts introduced in 1864 and 1871 granted protection to domestic refiners from imports of oil products, and domestic refiner Desmarais Frères constructed its first refinery in 1871. Yet, in a reversal, taxes were imposed on refined products in 1903 and 1908, lessening the difference between the value of imports of crude and refined products. This led to a rise in product imports. In 1900, ninety-three percent of oil products were refined domestically, but this fell to forty percent in 1908, and seven percent by 1927.[5]

Business competition focused on how to distribute the profits within this growing market. Imports from the United States were initially managed by French refiners, which formed a trade association in 1892 and contracted for supplies. The reduction of Russian oil meant that by 1918 US oil grew to make up almost ninety percent of domestic demand. This improved the commercial position of Standard Oil (New Jersey).[6]

The interests of Standard Oil were not simply focused on profits in the French market. Instead, they – along with other firms with international interests – sought to control the emergence of new centers of oil production in the Middle East and elsewhere in order to maintain a floor on prices. Negotiations between business and government over how to govern the domestic market were therefore influenced by the attempts of Standard Oil, Royal Dutch Shell, and others, to control global trade in oil from well to pump, including by expanding into the domestic market rather than simply contracting with local firms.

The organization of the French market mattered to the majors in seeking to achieve this goal. It also mattered to France's independent refiners, banking interests, and the government. One attempt to regulate the distribution of market share aimed to counterbalance the

[4] The first registered import of oil was three barrels of US oil by way of Liverpool. The first direct shipment was sixty barrels from New York into Le Havre. Romanian oil was first imported in 1903. Customs data on the mix of crude oil and oil product imports is available in Morgan Le Dez, "France's Oil Sources and Supply Networks (1861–1950)," in *Oil Producing Countries and Oil Companies*, ed. Alain Beltran (Brussels: P.E. Lang, 2011), 15–32.

[5] Christian Rouxel, *D'Azur à Total: Desmarais Frères, le premier grand pétrolier français* (Paris: Editions Drivers, 2007).

[6] Gregory P. Nowell puts non-Russian and US oil at fifteen percent of total imports. Nowell, *Mercantile States and the World Oil Cartel, 1900–1939* (Ithaca: Cornell University Press, 1994), 95–6.

dominance of Standard Oil (New Jersey) by improving the commercial position of Royal Dutch Shell. The Société pour l'exploitation des pétroles (SEP) was established in 1919 and given rights to exploit the share of Mesopotamian oil allocated to France through the 1920 Treaty of San Remo.[7] The SEP favored the interests of Royal Dutch Shell, which was given forty-nine percent of its equity. A broad range of oil, banking, and industrial interests were also allocated shares.[8] Political interest was therefore inseparable from private gain; both business and government sought to shape the distribution of market share domestically, firms for reasons of profit, and policymakers in order to moderate the market power of Standard Oil interests.

The SEP initiative failed in the face of opposition from Standard Oil, which halted deliveries to France in response to the favoritism shown to Royal Dutch Shell. Business and government interests then renegotiated how to govern the domestic oil market and the structure of French interests internationally. One focus of this effort was the development of domestic resources, as in Japan. In 1922, for example, the domestic mining law was revised to clarify property rights issues associated with exploring for and producing oil, and subsidies were established for exploration within national borders.[9] Yet, also like Japan, these efforts were disappointing.

The industrial compact used to govern the French oil market was negotiated in the decade after the failure of the SEP initiative. The results ended up lasting for six decades. In 1925 parties of the left sponsored the creation of the Office National des Combustibles Liquides (ONCL), which was given some ability to regulate

[7] The share represented the assets of the Deutsche Bank, which had been seized by the British government at the outbreak of World War I given that TPC was registered in London. For one account of these negotiations see Eric D. K. Melby, *Oil and the International System: The Case of France, 1918–1969* (New York: Arno Press, 1981). Individuals such as Henry Bérenger had an important influence over negotiations over French interests in Mesopotamia. Nowell, *Mercantile States and the World Oil Cartel*, 125–6. Nayberg agrees that Bérenger was important in the formulation of the government's position. Roberto Nayberg, "La politique française du pétrole à l'issue de la première guerre mondiale: perspectives et solutions," *Guerres mondiales et conflits contemporains* 4 (2006): 111–33.

[8] Nowell, *Mercantile States and the World Oil Cartel*, 134.

[9] Office National des Combustibles Liquides, *Annales de l'Office National des Combustibles Liquides Vol. 1* (Paris: Société d'éditions techniques, 1926), 27; on the incentives for domestic exploration see 25–7.

domestic industry. The ONCL regulated a system of stockpiling requiring four months national consumption of gasoline and five months of crude oil by 1926.[10] Between 1926 and 1928 the government then claimed the right to license imports of crude oil and products, and delegated this right to firms.[11] Quotas were administered through crude import licenses, which were issued for twenty years, and refined product licenses, which were issued for three years.[12] The first group of recipients was allocated market share in April 1931. This group included the majors, along with a newly created French public–private partnership – the Compagnie Française des Pétroles (CFP), described below. Importers were required to maintain reserve stocks within France, have senior managers of French nationality, and obtain prior authorization for increasing capital. Although they were not required to own the physical storage, there were few options for leasing available, creating a further barrier to market entry. Other components of the law established the legal authority for the government to force private-sector firms to submit to national supply contracts.[13]

The most important instrument used to mediate the interests between the majors, local firms, government, and banks was a public–private partnership with equity divided between the major business interests and the government. The CFP was created in 1924 with equity distributed to Standard Oil, Royal Dutch Shell, French banks, and domestic

[10] For firms with domestic storage this did not represent a high barrier to entry because those firms kept inventory for commercial reasons. For storage figures see Office National des Combustibles Liquides, *Annales de l'Office National des Combustibles Liquides Vol. 1*, 44.

[11] This was opposed by Royal Dutch Shell, which preferred cartelization through the cooperation of business rather than a system of formal quotas. Nowell, *Mercantile States and the World Oil Cartel*, 168. An alternative scheme to create a national monopoly was also opposed. The 1925 law can be found in Office National des Combustibles Liquides, *Annales de l'Office National des Combustibles Liquides Vol. 1*, 154–7.

[12] The law also required importers to maintain stocks, and Article 8 gave the government the authority to subsidize investments by French firms in new refineries built domestically.

[13] The importation of refined products was limited to gasoline, lubricants, and heavy and gas oils, with fuel oils, paraffins, and other products remaining outside the licensing system. Dominique Finon, "French Energy Policy: The Effectiveness and Limitations of Colbertism," in *European Energy Policies in a Changing Environment*, ed. Francis McGowan (Heidelberg, Germany: Physica Verlag, 1996), 24.

independent refiners such as Desmarais Frères. CFP was allotted a 23.75 percent stake in the TPC/IPC in 1927. In 1931 the government acquired a thirty-five percent stake in the firm and forty percent of voting rights. At the time of its establishment Poincaré claimed a national interest for the firm, yet the final structure of the CFP was also useful for the majors and other participants.[14]

The CFP public–private partnership extended into domestic refining through the Compagnie Française de Raffinage (CFR). This relationship between CFP and CFR was cemented by the shareholding structure: CFP was allotted fifty-five percent of the shares in CFR, with thirty-five percent distributed to the five largest French refining firms. ONCL provided capital to aid in CFR's start-up by taking a ten percent equity stake.[15] CFR also received a mandate to refine twenty-five percent of domestic demand, taking market share away from the independent refiners even as they received a share of equity in the firm.[16] By 1936 CFR supplied almost twenty percent of domestic demand, although it also retained agreements with independent marketers that stopped it selling products domestically other than to large industrial users in cement, electricity, and other industries.[17]

Lastly, retail prices were fixed, along with the margins for storage and refining. These were adjusted on an annual basis through negotiations between firms and the government, and in response to inflation and changes in the cost of crude. Prices were set in order to ensure that France remained profitable for refiners to invest in, with the goal of encouraging the growth of the French refining sector.[18] The success of

14 Shares taken by other firms were: British Petroleum (23.75 percent), Royal Dutch Shell (23.75), Near East Development (Standard New Jersey and Socony Mobil) (23.75), Partex Gulbenkian (5).

15 See Nowell, *Mercantile States and the World Oil Cartel*, 177 and 207 for the division of shares in CFP and CFR.

16 Kuisel recounts the difficult process of passing enabling legislation through parliament. Richard F. Kuisel, *Ernest Mercier: French Technocrat* (Berkeley, CA: University of California Press, 1967), 38–44.

17 Leslie E. Grayson, *National Oil Companies* (Chichester: John Wiley and Sons, 1981), 50. This harmed the interests of independent refiner Desmarais Frères. See Rouxel, *D'Azur à Total*.

18 The setting of domestic product prices so as to afford greater profitability for firms is noted by J.H. Bamberg in his history of British Petroleum. See J.H. Bamberg, *British Petroleum and Global Oil 1950–1975: The Challenge of Nationalism* (London: Cambridge University Press, 2000), 240. See also Andre Philippon, "The French Example: The 1928 Laws," in *A Comparative History*

the law can be seen in the establishment of fifteen refineries in France after it was passed, with the two largest owned by CFR. Refinery capacity increased from 300,000 tons per year in 1927 to eight million tons per year in 1938. Imports of refined products as a ratio of total consumption fell from ninety-four percent in 1929 to eighteen percent in 1938.[19]

Negotiating the industrial compact concluded between business interests and the government was contentious. Repeated attempts were made to cartelize the domestic market with different distributions of market share, and to give precedence to the oil produced by different business groups' interests. There were also divisions within the government, and between firms, with some siding with the interests of Royal Dutch Shell, some with Standard Oil, and others proposing a different set of solutions. Yet, by the end of this process, the French market moved away from a structure in which Standard Oil supplied refined products to domestic marketers, to one in which the interests of Standard Oil, Royal Dutch Shell, French banking interests, the domestic refiners, and the government were coordinated through a private–public partnership that allocated market share between these interests, and operated in refining, distribution, and marketing within the French market. Interests were further balanced through a system of import quotas for crude oil and oil products, which governed CFP market share and the share of other market participants. Internationally, CFP operated in exploration and development activities. Together this proved a remarkably durable form of market governance.

The compact used to regulate market share was reconstituted after the liberation of France. Britain seized CFP's holdings in the Middle East in July 1940 under the Trading with the Enemy Act, but they were not dissolved.[20] Oil import authorizations were suspended in 1941, but new dates for authorization were established in 1948. The twenty-year licenses for crude oil imports were reduced to thirteen years by decree

of National Oil Companies, Alain Beltran (ed.), (Brussels, Belgium: Peter Lang, 2010), 34. Elf was also given the right to revenues from production in the Lacq field without being required to provide tax to state coffers. Harvey B. Feigenbaum, *The Politics of Public Enterprise: Oil and the French State* (Princeton University Press, 1985).

19 Mohamed Sassi, "The Emergence of a French Oil Industry between the Two Wars," *Business and Economic History On-Line* (2003): 19–20.

20 Melby, *Oil and the International System*, 175.

in 1950, with quotas set for gasoline and heavy oil. When they came up for renewal once again in 1963 the licensing period was reduced to ten years, meaning renewals were carried out in 1973 and 1983. The implied market share of French refining and distribution firms stood at 49.6 percent following the distribution of quotas in 1950.

One important change to the equilibrium was the establishment of a second corporate entity. In 1941 the Société nationale des pétroles (SNPA) was created and given the right to explore within France in the Aquitaine region, with the government holding fifty percent of its equity and CFP retaining fourteen percent. In 1945 the government's share in SNPA was transferred to the Bureau de recherche de pétrole (BRP), as were the rights to an oil exploration firm based in Morocco, and later in Tunisia, Gabon, and Madagascar. The BRP also invested in Algeria through the Société national, des recherche et d'exploitation de pétrole en Algérie, a joint venture with the Algerian government.

As with the creation of CFR in addition to CFP as a vehicle for negotiating the interests of business and government, a new subsidiary firm dedicated to refining was established in April 1960. The Union générale des pétroles (UGP) was dedicated to refining Franc Zone crude – with BRP as the majority shareholder.[21] UGP established a presence in the French market by purchasing the refinery and distribution operations owned by Caltex, and was also given preferential treatment in the allocation of import licenses, gaining a market share of 14.6 percent as a result. By 1965 French firms were allocated 61.3 percent of the domestic refining market through refining authorizations, compared to 49.6 percent in 1951.[22] Franc Zone oil also helped increase the position of French firms. CFP began its exploratory work outside the Middle East for the first time in 1948, in the Algerian Sahara.[23] Its first big find was made in 1956, with production commencing in 1958. Finally, firms were given a tax exemption of fifty percent of profits invested in exploration within France and its colonies in 1953, and then globally from 1963.

[21] This was opposed by management of the Compagnie Française des Pétroles. Alain Beltran, *Elf Aquitaine: From its Origins to the Eighties* (Paris: Elf Aquitaine, 1999).

[22] Melby, *Oil and the International System*, 297.

[23] See Samir Saul, "SN REPAL, CFP, and Oil Paid in Francs," in *A Comparative History of National Oil Companies*, ed. Alain Beltran (Brussels, Belgium: Peter Lang, 2010), 93–123.

Postwar support for firms operating in the oil market echoed a consensus within government in favor of sectoral planning across the broader economy. In many cases the industrial strategy sought to shape the terms of competition in favor of domestic firms, and to support domestic firms in sectors that lacked a national industrial firm. French postwar oil policy was therefore mirrored in the strategy employed in steel, textiles, telecommunications, and other sectors.[24] At the core of the planning process were the Ministries of Industry and Finance. The latter was central to the allocation of credit, including the regulation of public agencies charged with providing long-term credit to the agricultural and industrial sectors. The Ministry of Industry, on the other hand, incorporated the interests of industrial sectors while implementing sectoral plans. In the postwar period, the chief regulatory agency remained the Ministry of Industry, which retained legal authority for monitoring the national oil companies and issuing licenses for the importation of oil and oil products. The Ministry of Industry was responsible for coordinating with industry and, as a line ministry, had an interest in supporting its clients in the form of national firms operating in the sector.[25]

The creation of UGP alongside BRP, as well as the existence of multiple subsidiaries under each, led to a complex set of shareholding structures and a fragmented firm structure. These were integrated in 1965 into the Entreprise de recherches et d'activités pétrolières (ERAP). The Elf trademark emerged in 1967 and subsumed all firms other than SNPA. SNPA was then merged with ERAP in 1976, with the new entity called Société nationale Elf Aquitaine, leading to the creation of the two national champions in oil: the CFP and Elf-ERAP.

24 For summaries and analysis of French industrial policy see Peter A. Hall, *Governing the Economy: The Politics of State Intervention in Britain and France* (New York: Oxford University Press, 1986); William J. Adams and Christian Stoffaës (eds.), *French Industrial Policy* (Washington, DC: The Brookings Institution, 1986); Saul Estrin and Peter Holmes, "French Planning and Industrial Policy," *Journal of Public Policy* 3 (1983): 131–48; John H. McArthur and Bruce R. Scott, *Industrial Planning in France* (Cambridge, MA: Harvard College, 1969); Jonah D. Levy, *Tocqueville's Revenge: State, Society, and Economy in Contemporary France* (Cambridge, MA: Harvard University Press, 1999), 18; Elie Cohen, "Industrial Policies in France: The Old and the New," *Journal of Industry, Competition and Trade* 7 (2007): 215.

25 Gilles Bellec, Ministry of Industry, interview with author, Paris, France, July 18, 2008. See also N.J.D. Lucas, "The Role of Institutional Relationships in French Energy Policy," *International Relations* 5 (1977): 95.

French policy entered the 1970s, therefore, with market share within the domestic market governed through a licensing law, along with two groups with government holdings that held approximately twenty-three percent of the domestic market share each across petroleum products. These firms – which later operated under the umbrellas of Elf-ERAP and CFP/CFR – owned productive capacity internationally that was equivalent to total French consumption.

The emergence of two vertically integrated firms with substantial domestic market share and international holdings cannot be attributed to the relative abundance of resources in France or its colonies. While French firms gained preferential access to oil reserves internationally in Algeria and elsewhere, the commercial exploitation of these reserves could have taken a variety of forms, including the continued reliance on the majors. The Japanese case also shows that access to reserves internationally does not determine the form of corporate organization. The emergence of these firms also cannot be attributed solely to interests of the government. Evidence shows that the 1928 legislation, as well as the creation of CFP itself, benefitted the majors in addition to domestic business interests, and that firms, including the majors, played a crucial role in negotiating the structure of oil market governance. Nevertheless, the outcome of this process of bargaining between government and business interests was the emergence of new firms that enjoyed substantial market share in refining and sales, and were also integrated into the upstream.

The role of Europe

I noted in Chapter 3 that the influence of European organizations offers an important alternative explanation for the changes in oil market governance we see in France. It is therefore necessary to identify European preferences towards oil markets prior to 1980. I focus in particular on the European Commission, which was important in driving proposals in the energy sector within Europe.[26] Evidence shows that the European Commission consistently supported reducing barriers to trade and investment between member countries, with the goal of completing the European single market. This extends to the energy sector, where the history of the engagement of European institutions

[26] Nicholas Jabko, *Playing the Market: A Political Strategy for Uniting Europe, 1985–2005* (Ithaca: Cornell University Press, 2005).

in the oil policies of national governments, as well as in gas and electricity, is largely defined by the attempt to realize this goal.

Evidence also suggests, however, that the Commission's ability to affect member countries' energy policies was limited; European agreements in fuels other than coal and nuclear energy continued to be governed by national, rather than supranational rules, meaning general rules governing trade and investment needed to be extended to cover the oil sector. Preliminary work by the Commission led the European Council to adopt a protocol covering energy issues in 1964. The protocol stated little more than a general desire on behalf of member states to harmonize national regulations in oil and natural gas, and it committed national governments only to conduct ongoing consultations with the European Commission.[27]

The lack of progress was implicitly recognized in the first attempt by the Commission to define a set of European goals for national energy policies across the member states. Released by the European Commission in 1968, the "First Guidelines for a Community Energy Policy" noted that

> measures at national level are leading to a gradual disintegration of the Community's energy economy ... and increasingly nationalist supply and marketing policies ... This dangerous trend can only be changed by a Community energy policy which fully integrates the energy sector into the common market.[28]

It argued that the guiding principle for a European energy community policy must be competition with the goal of lowering costs for consumers, with a common customs tariff for oil products, harmonization of product quality and technical standards that act as barriers to trade, and the abolition of cross-border barriers to investment.[29]

[27] Protocol of Agreement on Energy Problems, Reached Between the Governments of the Member States of the European Communities at the 94th Meeting of the Special Council of Ministers of the European Coal and Steel Community, April 21, 1964.

[28] Commission of the European Communities, "First Guidelines for a Community Energy Policy. Memorandum Presented by the Commission to the Council," Bulletin of the European Communities, Supplement to No. 12–1968, COM (68) 1040 final, December 18, 1968, 5.

[29] European Commission, "First Guidelines for a Community Energy Policy," 7. The recommendations on free movement of goods and freedom of establishment are on 11–12.

Each of these statements of principle was contrary to existing French institutions, which gave authority to the government to regulate the domestic refining market in order to discriminate in favor of French national oil champions.[30]

Oil market volatility increased, rather than weakened, the hold of national governments over energy policymaking. Recommendations from the European Commission were designed to enable better information to be gathered at the European level about imports and exports to and from member countries, and consumption of crude oil and petroleum product within them. The incompatibility of European goals with French national policy is revealed in the initial proposals made by the Commission regarding how to insure against oil supply risks. Guidelines prepared by the Commission in 1973 argued that major importers should stop firms from overbidding for contracts and driving up prices, and should harmonize measures for responding to supply crises. The Commission went further, however, proposing that the Council develop common rules for importing oil and oil products in order to "optimize the structure of the Community's oil industry while upholding competition among the various types of supplying companies."[31] The Council did not, however, produce a decision to this effect.[32]

In sum, the authority of the Commission sought to extend the principles of freedom of movement of goods, and of establishment, to the energy sector. Yet, while European authority extended into stockpiling measures and information-gathering, national governments refused to lower trade and regulatory barriers to trade in oil or oil products, and also did not succeed in harmonizing regulatory approaches to national firms operating in oil. The inability of the Commission to extend its mandate was evident in its communication to the European Council in October 1981:

[30] A 1968 European Council Directive required member states to stockpile the equivalent of sixty-five days of oil consumption, increased to ninety days under a subsequent 1972 directive. Neither made recommendations about market opening. There were also no provisions made for oil sharing or the joint release of stocks, meaning it did not promote the pooling of sovereignty over the management of oil stocks. The relevant directives are 68/414/EEC and 72/425/EEC.

[31] Commission of the European Communities, "Guidelines and Priority Actions under the Community Energy Policy," SEC (73) 1481 final, April 19, 1973.

[32] For a summary of the 1973 Council meeting see Guy de Carmoy, "The European Alternatives: The Community Energy Policy: Oil," *Government and Opposition* 12 (1977): 411–29.

In the course of recent years the European Council has repeatedly declared the need for the Community to face up to the energy challenge ... But it has not led to the implementation of an overall strategy comprising action by the Community, member states and producers and consumers. The inadequacy and inconsistency of the action taken in the wake of these expressions of political will can only be deplored.[33]

Transforming oil market governance in France

The laws and regulations used to govern France's oil markets were transformed in the two decades following the oil shocks. Years of contentious negotiations between business interests and the government had produced an equilibrium that distributed market share between Elf Aquitaine, CFP/Total, and other market participants. Market shares were regulated through a licensing system that conferred authority to government to intervene in the domestic market in order to increase the market share of French firms, and price controls helped create a domestic refining industry. Yet from 1982 price controls were relaxed, and were finally abandoned in 1986. The government also relinquished its holdings of CFP/Total and Elf Aquitaine, as well as the golden share and representation on the board. A revised version of the law governing the oil sector was passed on December 31, 1992, and it eliminated discrimination in favor of national firms. By the time this process was complete, the laws governing the French oil sector that had remained in place for almost seventy years, other than during the interlude of war, were transformed.

The transformation of the institutions of oil market governance occurred in three stages. The system of fixed prices was the first to be abandoned. A new pricing mechanism was negotiated between 1981 and 1982 that more closely linked domestic prices with the international market, and was finally abandoned in 1985–6 along with fixed margins for refining and storage. The system of import licensing was abandoned next, followed by privatization. The most significant reduction in state holdings in Total came in 1992, when

[33] Commission of the European Communities, "The Development of an Energy Strategy for the Community: Communication from the Commission to the Council. Mandate of 30 May 1980," COM (81) 540 final, October 2, 1981, Bulletin of the European Communities, Supplement 4/81, 7–20.

the share held by the state fell from 31.7 percent of outstanding shares, to 5.4 percent. The firm became a wholly commercial entity as part of its merger with Elf Aquitaine in 2000. Shareholdings in Elf Aquitaine fell from sixty-six percent in 1975 to 51.6 percent in 1992, thirteen percent in 1994, and finally zero percent in 1996. In 2000, the government also relinquished its ability to veto any attempted takeover by another firm if deemed not to be in the national interest.[34]

The new law introduced in 1992 did not ignore the problem of security of supply, but discarded the goal of ensuring French security of petroleum supplies through shaping flows of crude and crude products in favor of French firms; the first article of the law demonstrated the difference in approach from the law of 1928: "Article 1: Subject to compliance with the provisions of this Act, the import and export, processing, transportation, storage and distribution of crude oil and petroleum products can be conducted freely."[35] Instead of intervention under normal market conditions, the maintenance of strategic stocks became the focus of policy. This was, in other words, a choice to bring France in line with multilateral solutions to the security of supply problem in the petroleum sector: free trade, harmonized regulations, and strategic stocks maintained for emergency purposes. The new law also required firms operating in the petroleum sector to respond to government requests for information about import and export plans for crude oil and petroleum products.

In Chapter 3 I argued that the interests and strategies of firms were crucial in shaping policy outcomes in oil markets. I also claimed that firm characteristics shaped the strategies they adopted towards government. The institutions governing the French oil market had long

[34] There is one caveat to the broad transformation of French national institutions. Article 8 of the new law required refiners to notify the government of plans to build or scrap refining installations one month prior to the implementation of such a plan, and conferred to the government the authority to reject such a change if it determined that it might serve to disrupt the supply of petroleum products within France. The government also retained the ability to sanction firms failing to comply with an order by penalizing them with fines of up to ten million francs. Article 8, law no. 92–1443. Interviews suggest this article is designed to allow the government to maintain veto power over changes in refinery structure for reasons of local employment and local supply conditions.

[35] Law no. 92–1443.

served the interests of French firms. How then, can we explain this change?

In this section I argue that the diversification and scale of Elf Aquitaine and CFP/Total predisposed them to support liberalization and privatization in the French petroleum market. In the long run, the shift in market structure towards greater diversification, and opportunities for exploiting reserves internationally, led firm management to pursue growth in new markets in crude production and the marketing of products. This reduced their interest in fighting to protect market share in the domestic French market. General tax policies remained important because firm headquarters were located on French soil, but policies designed to enhance firm control over the domestic French market became unimportant to firms. Indeed, to the extent they hindered growth, firms supported weakening the institutions of national control. French firms actively supported price liberalization, for example, a process that was driven forward by the entry of new market entrants into the oil market that took advantage of price flexibility in order to extend their market share.

In Chapter 3 I noted that oil market governance represents a bargained outcome between business interests and the government. I therefore also describe below how the strategies of policymakers changed in response to the transformation of the international oil market, and because of the problems with inflation and government indebtedness.

Markets and firm strategy

Strong demand growth ensured firms prospered. Data drawn from the annual reports of Elf Aquitaine and CFP/Total, and their predecessors, show that both firms took advantage of increased opportunities internationally by diversifying outside the Franc Zone and the Middle East, both in refining and oil production. This served to reduce the importance of the domestic market to the firms over time.

In the upstream, Elf-ERAP expanded internationally in crude as well as in natural gas, beginning in the 1960s. Spending in oil exploration grew from forty-four million francs in 1961 to 378 million in 1969. By 1964 about two-thirds of all spending in exploration (oil and gas) was located outside France, in Africa, North America,

Australia, and New Zealand,[36] a shift noted by the chairman in his address to shareholders for fiscal year 1968.[37] Exploring for oil is risky, yet many of these investments struck oil in commercially exploitable amounts; spending on the oil production phase also increased almost forty percent between 1963 and 1969 in nominal terms, rising from seventy-nine million to 207 million French francs. Annual oil production also increased from 603,000 barrels in 1960 to 8.3 million barrels in 1969. In the same year the chairman noted that the "first months of 1969 have clearly shown that the future of SNPA does not lie only in France."[38] True to this assessment, by 1971 production expanded across North Africa to include Tunisia, Libya, the Congo, and Nigeria, as well as Norway in the North Sea, and by 1980 production also included the United States, Iran, and Cameroon. In 1990 the geographic spread of firm production spread across fifteen countries globally, including Angola, Colombia, Oman, the Netherlands, the United Kingdom, and Italy.[39]

CFP/Total similarly began to increase crude production internationally. Data show that in 1969 production focused on Iran, Iraq, and Algeria, but this grew to include the North Sea, Indonesia, and North America by 1976.[40] Capital investment also focused on expanding refining operations internationally in order to secure outlets, as CFR began to produce more crude than its refining arm could refine. CFP/Total secured its first share in distribution through a company in West Africa, where it was not constrained in entering the downstream sector by existing commercial agreements. By 1970 sales of petroleum

[36] Société nationale des pétroles d'Aquitaine, *Rapport Annuel 1964* (Paris: SNPA, 1965), 8–11. It is worth noting that some of this test drilling ended in failure, with the exceptions of the discoveries at Rainbow Lake in Canada, and later at the El Gassi field in Algeria. Both served to increase commercial crude production by the firm.

[37] "Extract from the Chairman's Address," in Société nationale des pétroles d'Aquitaine, *Rapport Annuel 1968* (Paris: SNPA, 1969), 4. Data calculated from Société nationale des pétroles d'Aquitaine, *Rapport Annuel* (Paris: SNPA, various years). Figures are for both oil and gas exploration and production. The chairman of SNPA recognized the potential for this expansion in the Annual General Meeting of May 26, 1965.

[38] Société nationale des pétroles d'Aquitaine, *Rapport Annuel 1968*, 4.

[39] Data drawn from Société nationale Elf Aquitaine/Elf Aquitaine, *Rapport Annuel*, 1966–90.

[40] Compagnie Française des Pétroles, *Rapport Annuel*, 1969–76. Data reported on a regional rather than country basis in later years.

products were only marginally greater within France, standing at 21,497 million tons compared to 19,050 tons internationally.

This expansion of CFP/Total operations was not only organic. The firm acquired a number of independent French firms throughout the 1950s and 1960s. The appearance of the Total brand in 1953, and its adoption in France in 1957, was coordinated by CFR. It enabled some twenty independent French marketers to sell under a single brand, and by 1970 the last two significant independent refiners were acquired; Desmarais Frères was acquired by CFR in 1966, and Elf-ERAP acquired ANTAR in 1970.[41] These changes were facilitated by the diversification of opportunities to produce oil for firms other than the majors. The success of the diversification of CFP/Total and Elf-ERAP shows both firms were in a good position to take advantage of this change.

In Chapter 2 I noted that oil price volatility was a second consequence of a shift towards a more open international market for oil. In the short term, French firms incurred substantial losses from the nationalizations that precipitated this change. Firm holdings in Algeria were secured during negotiations over Algerian independence: the French and Algerian governments reached an agreement in 1965 to change the legal structure for exploration and production work, with a joint venture established called the Société Petroliere Française en Algérie (Sopefal) between ERAP and the Algerian firm Sonatrach.[42] In 1971, however, the Algerian government nationalized gas and oil assets and signed new agreements with CFP/Total and Elf-ERAP on less accommodative terms. The El Gassi field dominated Elf-ERAP oil production in Algeria following the completion of the Sonatrach pipeline in March 1966, with annual production reaching 12.5 million barrels in 1966.[43] After Algerian independence it remained the second largest center of production after Gabon, but by 1976 Algerian production was wholly lost to the firm.[44]

CFP/Total production levels were also harmed by nationalizations. Algeria, which dominated production along with Iraq, fell by more

[41] Grayson, *National Oil Companies*, 60.

[42] French companies were then required to give any exploration permits they held to the joint firm in return for a payment of the book value of the asset.

[43] Data drawn from Société nationale des pétroles d'Aquitaine, *Rapport Annuel 1966* (Paris: SNPA, 1967).

[44] Société nationale Elf Aquitaine/Elf Aquitaine, *Rapport Annuel*, various years.

than half. The operations of the IPC in Iraq, in which CFP/Total continued to hold equity, were nationalized in 1972, although CFP/Total assets in the south of Iraq were left untouched by the nationalizations through the Basrah Petroleum Company. By 1976 production in Iraq had also dropped by more than half from its peak in 1974. Total CFP/Total production fell by sixteen percent by 1976 from its peak in 1974.[45]

Rather than seek the protection of government, however, both firms responded by further diversifying operations. Management argued that the best response to the nationalization of some of their most productive assets lay in developing new sources of oil and in improving productivity. At the Annual General Meeting for the financial year 1979, the president of Elf argued, for example, that the problems of the 1970s demonstrated that access to equity oil was crucial in ensuring firm success, and made international exploration and trading the "absolute priority" within the firm.[46] The sale of crude internationally – which made an increasing contribution to Elf Aquitaine's profits – and the number of countries from which Elf Aquitaine produced crude grew, corroborating this increased importance of oil production within the firm.

Elf's diversification strategy extended to refining and marketing. Both Elf and CFP/Total entered the US market in order to extend the geographic scope of downstream operations.[47] Elf also made investments in the United Kingdom and Spain.[48] In its annual report of 1991 the company continued to define its goal of increasing reserves and internationalizing activities in refining.[49] This diversification strategy was mirrored by CFP/Total, which extended into refining and

[45] Figures taken from Compagnie Française des Pétroles, *Rapport Annuel*, various years.

[46] President's Address to Annual General Meeting of Shareholders, June 12, 1980.

[47] Staff reporter, "Elf Eagerly Courts Downstream Investment by Producing Countries," *International Petroleum Finance* 11 (1988): 1.

[48] Staff reporter, "Elf Chairman Eyes Profits, Not Sentiments," *Petroleum Intelligence Weekly* (September 30, 1991): 8; Staff reporter, "French Refiners Face Shakeout with Soaring Costs, Downturn," *Oil & Gas Journal* (September 20, 1993): 25.

[49] Elf Aquitaine, *Rapport Annuel 1991* (Paris: Elf Aquitaine, 1992), 2. The expansion was closely tied to the management strategy of CEO Loïk Le Floch-Prigent. It also was not without problems: the firm disposed of assets of six billion francs in value by the end of FY1994, partly in order to reduce debt. See Elf Aquitaine, *Rapport Annuel 1994* (Paris: Elf Aquitaine, 1995). It is

distribution outside France, and diversified investments upstream. Sales of refined petroleum products in France as a ratio of total sales fell from forty-one percent in 1983, to thirty-five percent in 1990, to nineteen percent in 2004, meaning that sales in the rest of the world grew from fifty-eight percent to eighty-one percent over the same period.[50] The share of sales in Europe, as a component of total sales outside France also increased over this period, and grew particularly rapidly following the entrance of the firm into Eastern Europe. The firm – now under the Total banner – reorganized business operations in 1990 in order to promote international operations. The reorganization of the business was made, in the words of the 1990 annual report, because "Total's basic refining and marketing operations now cover the whole of Europe, while its downstream activities are worldwide in their reach. The reorganization is designed to adjust to this new situation."[51] In refining it unified its management structure for Europe, meaning that it no longer had a national organization around the refining and marketing of petroleum products, but rather optimized operations across Europe. This reflected the poor performance of legacy refineries in France; the utilization rate in French refineries stood at just sixty percent, compared to close to ninety percent in the four refineries operated in Total Petroleum North America.[52]

All this meant that the French market mattered less to the firms. Sales of crude oil fell in France as a ratio of total sales. In 1983, for example, the Total group sold some nineteen million tons of crude in France, which increased to twenty-one million tons by 1990, an increase of nine percent. Sales to the rest of the world, on the other hand grew from twenty-four to forty-seven million tons over the same period.

worth noting that Elf Aquitaine owned assets across natural gas, sulfur/sulfuric acid, crude and products, petrochemicals, and later hygiene and healthcare. Sales were initially dominated by gas, with crude and products increasing in importance in the late 1960s and 1970s. Sales in the other business sectors Elf Aquitaine was involved in were gas (1975: twenty-seven percent), sulfur (seven percent), chemicals (thirteen percent), hygiene and health (nineteen percent), and coal (two percent). Data compiled from Elf Aquitaine, *Rapport Annuel*, various years.

50 Includes sales of gasoline, jet fuel, kerosene and gas oil, fuel and heating oils, lubricants, and other more minor products. Data here, and below, calculated by author from Total annual reports, various years.

51 Total, *Rapport Annuel 1990* (Paris: Total, 1991), 18.

52 Staff reporter, "Total-CFP," *Petroleum Economist* (July 1988): 266.

The effects of oil market volatility in refining and marketing accentuated this shift, as margins fell because of reduced demand for refined products and refinery overcapacity.[53] As in Japan, the rapid increase in oil prices had a significant and negative effect on the demand for refined petroleum products, both because of falling demand, and because of a shift in the composition of demand for electricity generation and other industrial processes away from oil. Between 1973 and 1975 French demand for crude fell by 330,000 barrels per day, equivalent to a fall of more than twelve percent from 1973 levels. Between 1979 and 1985 demand fell by almost 650,000 barrels per day, or a quarter of the level of demand in 1973. The total drop in demand represented 770,000 barrels of oil per day between 1973 and 1985, or some thirty percent of aggregate demand.[54] As a result the refining group of Elf Aquitaine rarely made a significant contribution to group profits.[55] The management's assessment was that "refining is an industry that is destined to decline, at least in volume terms."[56]

Firm strategy and oil market governance

Growing diversification undermined the usefulness of the existing laws and regulations governing the French oil market. Indeed, nationalization forced firms to actively lobby for dismantling them. Nationalizations removed control over crude price setting from French firms and handed it to the governments of the producer countries.[57] In the short term, the government stopped Elf Aquitaine and other firms from fully passing on the increased crude oil procurement costs to consumers during the 1973–4 crisis.[58] Arguments made by the management of Elf Aquitaine to allow for price increases went unheeded.

[53] Société nationale des pétroles d'Aquitaine, for example, predicted in its annual report for FY1972 that demand for petroleum products would continue to grow at an average of five percent annually for the next twenty years. See Société nationale des pétroles d'Aquitaine, *Rapport Annuel 1972* (Paris: SNPA, 1973), 5.

[54] Calculated from BP Annual Review of Statistics.

[55] Staff reporter, "Elf Aquitaine," *Petroleum Economist* (August 1989): 246.

[56] Société nationale Elf Aquitaine, *Rapport Annuel 1979* (Paris: SNPA, 1980), 2.

[57] The loss of control over price setting was noted by ERAP in its annual report. See Entreprise de recherches et d'activités pétrolières, *Rapport Annuel 1973* (Paris: ERAP, 1974).

[58] Société nationale des pétroles d'Aquitaine, *Rapport Annuel 1973* (Paris: SNPA, 1974), 3; Société nationale des pétroles d'Aquitaine, *Rapport Annuel 1975* (Paris: SNPA, 1976), 26.

Although eventually approved, they were inadequate to recoup the increased costs of crude procurement.[59]

This short-term insouciance of policymakers towards firm interests was compounded by the long-term effects of fixed prices under conditions of price volatility: with fluctuating crude oil prices – as well as a floating dollar–franc exchange rate – fixed product prices adjusted on an annual basis limited the ability of the firms to adjust end-user prices according to the changing costs of inputs. The liberalization of prices thus promised to enable firms to adjust prices in response to newly volatile crude input costs, partially arresting the losses the firms faced in refining.[60]

Firm complaints against the system were widespread. The president of the employers' association in the refining industry argued that the price of fuel was set too low, which was causing the industry losses increasing at a rate of one billion francs per month.[61] The advisor to President Mitterrand on energy issues fielded strong complaints from Esso-France and BP-France about fixed retail prices. Shell-France threatened to abandon operations in the country until a more flexible pricing mechanism was put in place.[62] Elf Aquitaine and CFP/Total also supported the loosening of prices: the firms lobbied the Directorate of Hydrocarbons within the Ministry of Industry, as well as the president's office, to reform the price-setting system so that adjustments were made on a monthly rather than an annual basis in order to better reflect the new reality of shifting prices for inputs.[63] Elf Aquitaine noted the problems the existing regulatory structure was causing for the performance of the refining sector at its Annual General Meeting of 1981:

The government is going to have to review this country's now outdated petroleum policy. It may opt for a genuinely protective dirigisme, enabling firms to finance industrial conversion, or alternatively it may opt for a laissez faire policy leaving them to shoulder both the risks, and the opportunities,

[59] Société nationale des pétroles d'Aquitaine, *Rapport Annuel 1973*, 3; Société nationale des pétroles d'Aquitaine, *Rapport Annuel 1974* (Paris: SNPA, 1975), 3.

[60] Senior policymaker in oil sector, interview with author, Paris, France, August 20, 2008.

[61] Staff reporter, "Pétrole: les raffineurs inquiets [Oil: The Concerns of Refiners]," *La Croix* (February 27, 1982): 7.

[62] Jean-Daniel Levi, Energy Advisor to President Mitterrand (1982–8), interview with author, Paris, France, July 8, 2009.

[63] Gilles Bellec, interview with author; Jean-Daniel Levi, interview with author; Societe nationale des petroles d'aquitaine, Rapport Annuel (Paris: Societe nationale des petroles d'aquitaine, 1973), 4.

of achieving this by their own efforts. The main priority is to break out of the existing hybrid system, which is gradually bringing about the demise of refining in France.[64]

Pressure from firms led to negotiations over how to adjust the price-setting system to protect firm profitability while limiting the potential for inflationary effects. Negotiations between the firms and the government began in 1982.[65] On April 15, the Ministry of Finance held a roundtable with representatives of the petroleum companies to discuss how to manage pricing. They determined, along with the Directorate of Hydrocarbons, that a new pricing mechanism should be introduced in order to link domestic prices to prices from the Rotterdam market and other European countries.[66] Whereas under the previous system prices were determined through a process of negotiations between firms and government representatives from the Ministry of Finance and the Ministry of Industry, with a complex series of adjustments made for regional differences, under the new formula it was agreed that the role of the Comité National des Prix should be limited to setting a ceiling price and fixing margins for distribution. Instead, under the new system product prices were automatically determined by the quoted prices in the Rotterdam market and the average price across Europe. The change was implemented over a three-month period to allow regulators to judge whether freer prices would prove too inflationary. Minister of Energy Edmond Hervé noted that the intention of the new system for determining prices was to balance the needs of producers and consumers by allowing refiners to gain a price similar to that of other markets, while also protecting consumers from excessive price volatility.[67]

The partial liberalization of prices was not a stable equilibrium. A price war followed the partial liberalization of prices, with supermarkets and other discount retailers offering rebates on petroleum prices at

[64] "Chairman's Address, Annual General Meeting of Shareholders," in Société nationale Elf Aquitaine, *Rapport Annuel 1980* (Paris: SNEA, 1981).

[65] Bruno Dethomas, "Les prix de l'énergie et les réformes [Energy Prices and Reforms]," *Le Monde* (March 3, 1982): 1.

[66] Bruno Dethomas, "La nouvelle politique pétrolière [The New Politics of Petroleum]," *Le Monde* (April 17, 1982): 31.

[67] Edmond Hevré, "Une politique pétrolière cohérente [A Coherent Oil Policy]," *Le Monde* (April 23, 1982): 1.

levels higher than the maximum rebate levels established in November 1982. All the major firms operating in the French market, except for Elf Aquitaine, attempted to match the rebates of the new market entrants, leading to both price destruction, and the loss of market share by the French firms and the majors. Elf Aquitaine then entered the rebate war, with Elf president Michel Pecqueur noting: "One must be pragmatic. There are rules, but they are not being implemented. We are not going to stand aside and let us incur losses due to a matter of principle."[68] Both Elf Aquitaine and CFP/Total therefore supported the shift to full price liberalization.[69]

Firms' preferences towards the relaxation of licensing requirements were equally positive.[70] The domestic market for refined petroleum products was not judged a growth sector by the management of the Elf Aquitaine and Total groups. Overcapacity, combined with the high costs of restructuring refining facilities due to changes in product structure, meant revenues from the refining of petroleum products fell while costs increased. In 1973 Elf Aquitaine owned shares in five French refineries (Ambes, Feyzin, Gargenville, Grandpuits, and Hauconcourt) and four others in Europe, but outside France (Reichstett, Albatross, Speyer, and Klarenthal). ANTAR, in which it had a significant stake, had shares in three French refineries (Donges, Vern-s-Seiche, and Valenciennes) and two outside France (Klarenthal and Herrlisheim).

68 Staff reporter, "La pratique des 'Super-Rabais' se généralise ['Super Discounting' is Generalized]," *Le Monde* (January 26, 1985): 19.

69 Elf chairman noted this in his address to shareholders, stating: "A major event in early 1985 was the removal of price controls, which was announced by the government at the end of January … We have no objection to this measure, and indeed the entire industry has been calling for it." Chairman's Address, Annual General Meeting of Shareholders, Elf Aquitaine, May 30, 1985. This is confirmed by Jean du Rusquec, Advisor to the CEO, Total, and former member of the Ministry of Finance: "Everyone was happy when oil prices were deregulated because the players realized that it was an old-fashioned system. Sometimes it was really a constraint because the finance minister delayed price increases that were justified because of shifts in the dollar–franc rate. The oil market was developing very quickly – it was no longer a fixed price; the share of spot became more important; there were daily fluctuations; so the administration of the fixed price mechanism was seen as problematic." Jean du Rusquec, Advisor to the CEO, Total, and former member of the Ministry of Finance, interview with author, Paris, France, July 23, 2008.

70 Gilles Bellec, interview with author.

Allocated market share hindered rather than helped firms adjust to the problem of overcapacity as they sought to shut down refinery capacity in the domestic market.[71] The long-term problems with domestic refining were noted by the Elf Aquitaine chairman in 1983 in remarks to shareholders: "The future of this sector looks bleak. World excess refining capacity (and particularly in Europe), and the awaited arrival of products from the Middle Eastern refineries, raises the question of just how much refining capacity France really needs."[72] Instead, firms were focused on market strategies centered on expanding beyond the French domestic market into Europe, the United States, and the Asia-Pacific, rather than pressuring policymakers to defend their market share domestically. Reflecting this, Elf Aquitaine began to optimize its refining operations across Europe rather than simply in France in the early 1980s, with a single program determining how supplies were distributed for the European market as a whole, rather than simply for the French domestic market.[73] The loss of mandated market share was therefore increasingly irrelevant, and was also balanced against control over market share afforded by the greater flexibility to organize refining.[74]

The transformation of the French institutions of national control was completed through the privatization of Elf Aquitaine and Total. Firms supported the change because of the constraints government ownership placed on their business operations, as with price liberalization and the removal of regulatory control over the domestic market. The government had less control over Total than Elf Aquitaine, with a maximum shareholding of thirty-five percent, compared to sixty-seven percent for Elf Aquitaine. Total welcomed privatization, as is made clear in management's assessment of the significance of the reduction in the government's share of control over the firm, and which is worth citing in full:

> The events of the past year mark the culmination of the transition of Total to a genuinely private, commercial entity as the Company revised its historical

[71] Roger Vielvoye, "Rapid Decline in Crude Oil Prices Buoys Profits for Refiners in Western Europe," *Oil & Gas Journal* (August 18, 1986): 25.

[72] Chairman's Address, Annual General Meeting of Shareholders, June 6, 1984.

[73] Olivier Abadie, Crude and Products trader, Elf Aquitaine (1990–5), interview with author, Paris, France, July 16, 2008.

[74] Edouard Thevenon, "Produits pétroliers: importations plus faciles [Petroleum Products: Importing is Easier]," *Le Figaro* (September 25, 1986): 18.

links to the French State. Management is fully aware of its responsibilities to the Group's new and enlarged shareholder base and of the need to make an even greater effort to improve performance, communication, and above all, to create value for our shareholders ... The transition from a State-controlled to a private commercial entity provides Total with access to a broader international investor base and more financial flexibility, while improving the liquidity of the market for Total's shares.[75]

The issue of government ownership was more significant for Elf Aquitaine. Like the Total group, managers viewed the idea of privatization positively. This was both because of the constraints government shareholding placed on the firm, and because the value of firm shares was understood by management to be undervalued relative to competitors because of government shareholdings. The increasing indebtedness of the government meant it was unable to play the role of a provider of capital, even as the firm was increasingly able to raise its own funds through international markets. This problem, according to managers, was compounded by the valuation of the stock relative to its competitors:

It was possible to issue new capital to the extent that the State was able to pay its share, but its share was 70 percent and the condition of the budget was not good. Development was strongly constrained by the fact that the State was not able to bring you new money. Total had the same constraint.[76]

Further, senior managers saw little benefit to retaining state ownership given that Elf Aquitaine was already listed in Paris and New York. This meant the firm was already required to meet the expectations of shareholders rather than the demands of the state.[77] The CEO of Elf Aquitaine argued to the government that market participants were scared by the intervention of the government, which was limiting their

[75] Total, *Rapport Annuel 1992* (Paris: Total, 1993), 6–7.

[76] Jean du Rusquec, interview with author. This is confirmed by Bruno Weymuller, who argued that the Treasury was unable to come up with funds to finance firm activities. Bruno Weymuller, Executive vice-president, Total (2000–8), interview with author, Paris, France, July 7, 2009.

[77] Gilles Darmois, Counselor to Martin Malvy, former State Secretary to the Minister of Industrial Redeployment and Foreign Trade, responsible for energy (1984–6), Advisor to the president of Elf Aquitaine (1991–4), interview with author, Paris, France, July 3, 2008.

ability to operate internationally; managers therefore wanted to reduce government ownership to ease their ability to access equity markets in order to fund its plans for international expansion.[78] The CEO noted in an interview that Elf was "already an international company confronted by international competition, and we seek the state's help neither financially nor politically. [Privatization] would only mean a greater involvement of private shareholders."[79]

The support of Elf Aquitaine for privatization was also driven by the divergence in the interests of policymakers and the firm. The government, through its rights as the largest shareholder in Elf Aquitaine, actively blocked an acquisition planned by Elf as part of its strategy of dealing with problems in the domestic refining and distribution market; in 1981 Elf wanted to enter the US market, but it was blocked by the Socialist government, which frustrated management.[80] Albin Chalandon, president of Elf Aquitaine, had already flagged his support for greater autonomy for Elf:

> I agree on the basic concept that was defined by the president of the Republic, namely that we are autonomous in our management and we must keep the initiative in developing our strategy, our policy over the medium and long term, but it must be an agreement between the government and us.[81]

This shift away from government as a provider of capital to one that was not useful to the firms, both because of its own fiscal crisis and because of the success of firms themselves, makes it unsurprising that the firms responded to the challenges of the nationalization of production and falling refining margins by redeploying firm resources towards greater international exposure, rather than calling for increased governmental support within the domestic market.[82] Elf's focus outside the French market is laid bare in a 1989 interview with the Elf Aquitaine

[78] Senior policymaker in oil sector, interview with author, Paris, France, August 20, 2008. See also Staff reporter, "Elf Eagerly Courts Downstream Investment by Producing Countries," *International Petroleum Finance* 11 (1988): 3.

[79] Petroleum Economist, May 1986, cited in Schmidt, *From State to Market?*, 147.

[80] Jean du Rusquec, interview with author.

[81] Staff reporter, "M. Chalandon au Grand Jury R.T.L," *Le Monde* (March 30, 1982): 42.

[82] I thank Jonah Levy for pointing out the importance of the fiscal crisis engulfing the French State, reversing its role as a provider of capital to firms.

CEO Loïk Le Floch-Prigent: "The general objective is very clear. Elf is to be one of the world's leading companies in oil and gas. And we will be ready to react to other good opportunities and do even more."[83] Looking back, the Advisor to the Total CEO and former Treasury representative summarized the forces leading to the dismantling of the regulations governing the French market as follows:

> The law was dismantled step by step. First, because it was no longer in line with the development of the oil and gas market. When it was promulgated there was not an international oil market. Oil was the business of the US, UK and companies largely under state control in the US. There was a relatively nationalistic approach, and this remained true after WWII. But in the 1970s the international oil market developed due to the development of North Sea oil and gas production. The way to guarantee security of supply was no longer the same.[84]

Shifting policymaker strategies

The diversification of operations by both Elf Aquitaine and Total made the laws and policies used to govern French oil increasingly irrelevant to firm strategy. Both firms focused on expanding outside French shores in order to increase equity holdings of oil, and growing their portfolio of downstream assets held internationally to reduce their reliance on the domestic market. Firm management therefore either actively encouraged, or acquiesced, to the liberalization of the French market.

What, though, of the government? How did it respond to firms that no longer sought protection or capital? Evidence shows that policymakers also became less interested in supplying support to firms, both

83 Staff reporter, "Elf Chairman Eyes Profits, Not Sentiments," 8.

84 Jean du Rusquec, interview with author. Cohen has noted that the shift of national champions away from supporting intervention by the state, and towards greater independence, occurred in telecommunications, aeronautics, and other sectors, also: "national champions began to see state intervention as a threat to their cash flow, to their own discretionary power and as an impediment to building international alliances. So the greater the success of the 'grand project,' the bolder the determination of globalized national champions to be freed from state dependence and to be judged according to market criteria." Cohen, "Industrial Policies in France," 217.

because of a shift in economic strategy, and because of the effects of oil market volatility on the existing system. The shift of policymakers towards supporting economic liberalization, including the privatization of broad swathes of French industry, has been taken up in detail elsewhere.[85] Parties of the left and right began to pursue economic growth and macroeconomic stability through a shift towards, and then away from *étatisme*. The Mitterrand-led presidency commenced a program of nationalization beginning in 1982, with the goal of improving firm performance. Some forty-six firms in all were nationalized, representing more than fifteen percent of all 1980 revenues for French industrial firms.[86] By the program's end, the government owned thirteen of the twenty largest French firms, and maintained controlling shares in others.[87] Nationalizations were carried out across the banking and industrial sectors, including sectors determined to be "strategic" to national economic performance, in which the goal was partly to reduce dependence on external markets.

The government was forced to reverse its expansionary policies. Mitterrand implemented what has been termed the "Great U-Turn," shifting towards fiscal austerity, reducing state subsidies for firms, and reducing public spending.[88] The change in government policy was married with privatization, as was the case in Japan in the 2000s. This process began in 1986. Under the direction of the Chirac government, the legislature passed laws enabling the divestiture of shares owned by the state of firms nationalized over the 1981–3 period, as well as firms in the banking and insurance sector. It was not uncontroversial. President Mitterrand refused to sign into law the bill enabling privatization at a meeting with the Council of Ministers on April 9, 1986.[89] Under a

[85] For a recent summary see Rawi Abdelal, *Capital Rules: The Construction of Global Finance* (Cambridge, MA: Harvard University Press, 2007). See also Schmidt, *From State to Market?*; Levy, "Economic Policy and Policy-Making"; Alain Guyomarch, "The Europeanization of Policy-Making," in *Developments in French Politics 2*, eds. Alain Guyomarch, Peter A. Hall, Jack Hayward, and Howard Machin (London: Palgrave Macmillan, 2001): 166–36.

[86] Pascal Dumontier and Claude Laurin, "The Financial Impact of the French Government's Nationalization/Privatization Strategy," *EFMA 2003 Helsinki Meetings* (2003). A review of the process behind nationalization can be found on 8–9.

[87] Schmidt, *From State to Market?*, 116–18.

[88] Schmidt, *From State to Market?*, 112.

[89] Staff reporter, "M. Mitterrand refuse la privatisation des sociétés nationalisées avant 1986," *Le Monde* (10 April, 1986): 1.

July 2, 1986, Act, the government was then given the authority to privatize firms other than those with a public service character. The law enabled the government to establish rules for determining the value of enterprises, the selling price, and other measures to facilitate the sales, including how the national interest would be protected. An August 6, 1986, law further specified the terms of privatization, including establishing a privatization commission to determine firm value and sales price, among other terms. This later extended privatizations to a number of firms not targeted by the initial wave of nationalizations.[90]

What of the oil sector? It is first worth noting that the oil sector was not subject to nationalization over the 1981–3 period. Both Elf Aquitaine and Total were privatized, however, as part of the shift to reduce direct ownership of firms by the state. In preparation for the privatization of Elf Aquitaine, the enabling law of August 6, 1986, was invoked to create a golden share (*action spécifique*), giving the government the right to veto share purchases deemed as not being in the national interest.[91] The government announced the sale of eleven percent of Elf Aquitaine capital on September 25, 1986. This was extended over time to include further share sales in both Elf Aquitaine and Total.

The general shift of parties of both the left and the right towards support for liberalization and privatization is necessary, but not sufficient, to explain the transformation of French policies of national control in the petroleum sector, however. This can be seen when comparing outcomes in the oil sector with the ongoing resistance to liberalization and privatization that continued in the other major sectors of France's energy industry – electricity and gas – throughout the 1980s and 1990s. Although all three were the focus of European Commission initiatives, for example, petroleum policies were adjusted quickly compared to those of electricity and gas, for which national policy dominated throughout the 1990s and 2000s.

The missing piece of the puzzle in explaining the transformation of French institutions to national control lies in the specific effects of the shift of the oil market structure on firms and policymakers. Changes in

90 On the technical procedures behind nationalization see Michel Berne and Gerard Pogorel, "Privatization Experiences in France," *CESifo Working Paper* (2004).

91 Staff reporter, "Privatisation: priorité à Saint-Gobain, action spécifique chez Elf," *Le Monde* (September 17, 1986): 45.

the international oil market already undermined the status quo by the time of the general shift in industrial strategy occurred. This meant that the drive to reduce the role of the state in the economy, supported by the proposed changes to petroleum regulations made by the European Commission in 1987, took place in an environment in which there was little interest among firms and policymakers to isolate the petroleum sector from the broader institutional changes being wrought across the French political economy. With firms no longer interested in defending the status quo, and the European Commission offering an alternative proposal for oil market governance, French policy in the petroleum sector was transformed.

The oil market introduced two sources of instability. First, it shifted the strategies of national firms that were at the center of the French strategy, as noted above, by stripping them of a portion of the oil they relied on for production, and by dampening demand and shifting the composition of demand domestically. For technocrats with delegated responsibility for ensuring the growth of the industry, the change in firm strategies meant they also had little reason to oppose liberalization and privatization.

Second, as in Japan, the poor performance of domestic refining and distribution opened the debate over the structure of domestic policy. Price variability, and falling demand, resulted in significant losses in refining for the firms, in the order of 700 million dollars in 1980–1 alone according to one estimate, and the performance of the industry through the 1970s was dismal.[92] These losses revealed a deeper problem with the system of fixed prices that governed the sector, as noted above: the need to adjust prices to reflect the new structure of the international oil market in which price-setting for crude oil was no longer internal to the international oil firms but was determined by firms and governments of producer countries. This variability in crude prices was accentuated by the pricing of oil in US dollars, meaning that the abandonment of the franc–dollar exchange rate exposed firms to exchange rate risk, given that revenues from sales within France were calculated in French francs, while firm procurement of crude oil from outside the Franc Zone were made in US dollars.

The Directorate of Hydrocarbons, which traditionally supported the interests of firms, followed them in supporting the loosening of

[92] Roger Vielvoye, "Jitters in France," *Oil & Gas Journal* (May 25, 1981): 122.

prices.[93] The position of the Ministry of Finance was more complicated. On the one hand, civil servants within the ministry understood the negative effects fixed prices were having on industry performance. On the other hand, fully liberalizing prices meant relinquishing an instrument that enabled them to influence sectoral outcomes, and also raised concerns about the potential for inflationary effects.[94]

The formula system was therefore a compromise. Firms with refining facilities continued to claim losses driven by the delay in price adjustment.[95] The government, on the other hand, was required to announce price increases on a monthly basis, a politically damaging task when gasoline prices were scheduled to increase. The decision to liberalize prices was therefore taken by Prime Minister Mauroy in mid-1984, but it was implemented only once prices began to fall.[96] This decision was supported by incoming Prime Minister Laurent Fabius, which is unsurprising considering that he was the minister with the industrial portfolio under the Mauroy government. The shift in price mechanism was also supported by incoming Minister for Economy and Finance Pierre Bérégovoy because of the positive effects it was likely to have on reducing inflationary pressures in the economy. Prices were therefore allowed to float freely across petroleum products from 1986. This marked the end of the system of fixed prices that had been designed to promote domestic industry by ensuring firm profitability within the French domestic market.

The second significant change focused on the delegated monopoly on trade in crude oil and petroleum products retained by the state. The law was replaced in 1992 by a new law in which the authority of the state to redistribute market share through imports and exports was abolished. Under the new law, the import and export of crude oil and oil products was fully liberalized. Although the enactment of the 1992 law represented the formal abolition of barriers to trade, the system

[93] Gilles Bellec, interview with author.

[94] Gilles Darmois, interview with author.

[95] There was not universal sympathy for the firms on this point, given increased profitability in crude production as a result of increased prices. In effect this meant that upstream operations were subsidizing refining and marketing downstream, with consumers still paying less than they otherwise would if prices had been fully liberalized.

[96] Gilles Darmois, interview with author. This is confirmed by another participant involved in decision-making. Senior policymaker in oil sector, interview with author, Paris, France, August 20, 2008.

was already not being applied as intended in law. While the directorate continued to require firms applying for A3 licenses to obtain the right to import petroleum products, the procedure had become increasingly redundant because of the national firms' success in establishing themselves in the domestic market, and because the firms themselves were becoming less reliant on the domestic market – a trend exacerbated by their responses to the oil shocks. The Director of Hydrocarbons at the time, Gilles Bellec, recounted the change in position of the ministry:

> In 1985–9 we were still dealing with the A3-A10 [licenses] although prices were liberalized. A3 companies came and asked for forms. We asked them for a supply plan for three years with three year contracts and how they were to get supplies, with 60 percent contract from one place, and 20 percent with Elf, and 20 percent from international market ... then the license was approved. That was the theory. But we were like priests not believing in God anymore. It was done; they would all present these supply plans. But we didn't believe anymore. The core thing was the three months [storage]. On the rest, they had good documentation and we studied it but less seriously.[97]

The disinterest of policymakers in defending the institutional status quo thus dovetailed with view of firms. The idea of loosening the issuing of licenses was raised by Minister of Economy and Finance Bérégovoy at the time of the shift to the full liberalization of prices in 1985–6. Bérégovoy supported relaxing the requirement that marketing firms purchase eighty percent of their products from licensed companies. Doing so, it was estimated, would allow firms to take advantage of cheaper products in the Middle East and elsewhere, thereby reducing prices in the domestic market and cutting inflation.[98] This proposed change was important. The system of import licensing lay at the heart of the law of 1928, and was the key to limiting price competition between firms in refining and marketing. The abolition of this system of licensing was, therefore, a significant step towards an open market for the supply of petroleum products.

[97] Gilles Bellec, interview with author. This was confirmed in another interview with a policymaker involved in the writing of the 1992 law, who noted that "The decision to write the 1992 was based on the fact that the 1928 law has no real effect." Senior policymaker in oil sector, interview with author, Paris, France, August 20, 2008.

[98] Jean du Rusquec, interview with author.

Not all policymakers agreed immediately with the change. State Secretary for Energy Malvy initially opposed the change on the grounds that the regulatory changes could negatively affect French firms struggling with problems of overcapacity. Over the long term, firms' positions in the domestic refining market were guaranteed through the requirement that eighty percent of products sold by distributors be obtained from within France. Removing this requirement promised to allow cheaper products into the French market, but also could have the effect of ceding national-flag firms' market share to refiners with capacity outside France, and to distribution firms in France with access to refined products produced outside the country.

The disinterest of the firms in defending the status quo meant baulking at this change was short lived, despite some opposition within the Directorate of Hydrocarbons, which was responsible for administering the licensing system.[99] As with prices, the directorate followed the lead of firms. The firms supported the adjustment Malvy implemented during his tenure; on August 15, 1982, he announced that this requirement could be fulfilled through refinery capacity throughout Europe, rather than within France alone. This suited the interests of Total and Elf Aquitaine, both of which were diversifying their refinery assets outside France and into Europe and elsewhere.

This change to allow European refinery capacity marked the first shift away from a sole focus on the French market. It also presaged the complete dismantling of the system of delegated monopoly. The so-called 80/20 rule – whereby refiners were expected to purchase eighty percent of their crude from holders of crude import licenses – was relaxed by Industry Minister Alain Madelin.[100] Madelin announced a number of measures, most importantly the immediate suspension of the rules of 80/20 and 90/10, which required importers of refined products to purchase from European refineries, and refiners to maintain refinery capacity equivalent to ninety percent of their domestic sales within France. He also extended by a period of six months the existing import authorizations of refined products to enable a new system to regulate trade to be determined.[101]

[99] Gilles Darmois, interview with author.

[100] This occurred under the cabinet of newly installed Prime Minister Jacques Chirac (1986–8). Veronique Maurus, "Le régime de la loi de 1928 sera maintenu," *Le Monde* (May 25, 1985): 21.

[101] Thevenon, "Produits pétroliers: importations plus faciles."

The final piece of the complete liberalization of the French oil market occurred through the privatization of France's two national champions: Elf Aquitaine and Total. For the firms themselves, managers favored full privatization, as noted above. The fact that shares were listed on the Paris and New York stock exchanges meant managers viewed them as already disciplined by the market.[102] Further, while the firms were able to rely on revenues to finance much of their investment, some in management thought that state holdings constrained their ability to raise capital, and also made the share price of the firm trade at discount because of investor fears of state intervention in firm planning.[103] And what of the government? The counterfactual is that the state would have retained its holdings in the national oil companies, absent the shift in firm sentiment. On this question there is evidence to support the contention that sector-specific factors determined outcomes: national firms in the other two major energy sectors with state holdings – natural gas and electricity – remained in state hands even after privatization and remained so into the 2000s.

There was also no support for retaining a state presence in the national oil firms other than through the *action spécifique*. The most likely opponent was the Directorate of Hydrocarbons, which managed sector-specific policies in the oil sector. It acquiesced to the change, however, both because of this support for privatization and because it did not have the power to veto the proposed privatization given support across the political spectrum for it.[104] The major controversy was therefore over the sale price for Elf Aquitaine shares, rather than whether the firm itself should be sold. The price determined by the privatization commission was less than the share price, leading some Socialists to oppose the conditions for sale.[105]

[102] Gilles Darmois, interview with author; Bruno Weymuller, interview with author.

[103] Jean du Rusquec, interview with author; Senior policymaker in oil sector, interview with author, Paris, France, August 20, 2008, Bruno Weymuller, interview with author.

[104] Senior policymaker in oil sector, interview with author.

[105] Veronique Maurus, "La vente par l'Etat d'une partie du capital d'Elf suscite une polémique," *Le Monde* (September 27, 1986): 35; Staff reporter, "Le parti socialiste s'indigne des conditions de vente par l'Etat de 11% du capital d'Elf Aquitaine," *Le Monde* (September 27, 1986): 40.

Summary

French businesses and policymakers negotiated an industrial compact with one another, and with the international oil major, which regulated shares in the domestic market through both licenses and through the creation of firms as vehicles for managing competition. This settlement was adjusted in the postwar period with the creation of a second firm, and this structure proved remarkably durable. Yet from 1980 the institutions used to govern the oil market were wholly dismantled. This stands in contrast to Japan, as we shall see in the next chapter, which made adjustments but continued to support domestic firms' position in the international oil market. The shift in French oil policy also stood in contrast to those in other energy markets, where the state retained shares in firms operating in said markets, most notably EDF and GdF.

In this chapter I argued that the transformation of French oil policy reflected the shift in strategy of firms. The industrial compact successfully created and then promoted the competitiveness of French firms, increasing their market share both domestically and internationally. This shaped the response of French firms to the shift in oil markets that occurred in the 1960s and 1970s. Both Elf Aquitaine and CFP/Total took advantage of the diversification of global production by expanding market share internationally into refining and exploration into the United States, other European countries, and elsewhere.

What of policymakers? The support of policymakers for the existing institutions governing oil markets also changed. The general transformation of the French model through privatization, price liberalization, and other such measures had an important effect on the timing of changes in the oil industry. Yet this general shift away from interventionism is not sufficient to explain outcomes in the oil sector, most notably because the effects of this movement were not evenly spread across industries. In fact, as noted above, the state continued to intervene heavily in the gas and electricity sectors, in contrast to oil.

To explain this outcome, therefore, we need to take into account the political costs to policymakers of maintaining the institutional equilibrium. This is most obvious in the case of fixed prices, which were of no benefit to policymakers, and began to be costly politically as crude prices began to decline. There was also little reason for the Ministry of Industry to seek to defend the institutional status quo, reflected in the law of 1928, when the firms were seeking to reduce capacity in the domestic market while focusing on expanding market share internationally. Further, increasing competition in the sale of oil products promised to help lower inflation, which was a major concern of policymakers at the time. Finally, in the case of privatization there was no one interested in retaining the status quo given the absence of firm demand for the state provision of capital retaining state ownership of the national oil firms.

Alternative explanation – the influence of Europe

I argued above that the transformation of France's institutions of intervention in the oil sector is best explained by firm and policymaker responses to the change in the structure of the international oil market. In this section I address an important alternative argument relevant to the French case: the influence of Europe. One way in which European integration affected the options available to French firms was by increasing the opportunities for cross-border investment.[106] Non-French Europe, for example, was a focus of CFP/Total from the 1980s, as it sought to diversify outside the national market. This can be seen from capital expenditure: in the 1980s spending by CFP/Total in European countries other than France stood at 2.1 billion French francs, compared to 2.5 billion within France. In the years between 1983 and 1994, capital expenditures within non-French Europe were slightly higher than spending within France itself at a mean of 30.1 percent of total spending, compared to 26.5 percent. Both were lower than spending in exploration and production internationally, however. Total capital expenditures spent outside Europe as a whole reached 5.5 billion francs in 1983 and averaged forty-three percent of capital expenditure over the 1983–94 period. Further, 75.5 percent of the firm's spending on capital in 1983 focused on the exploration, development, and production of oil, rather than in refining, which was more a focus of firm operations within Europe.[107]

106 I thank an anonymous reviewer for this point.

107 Data compiled by author from annual reports. The mean spending on exploration, development, and production between 1983 and 1987 stood at 74.1 percent. Elf Aquitaine annual reports do not break down data on capital expenditures to allow analysis of spending in non-France Europe.

The European Commission also began a concerted push to change French national institutions. By the time this occurred, however, the strategies of both firms and policymakers had already shifted to make them less interested in defending the national market against the encroachment of European regulations. This means that when the European Commission sought to complete an internal market for goods and services as envisioned in the Rome treaties, it found little opposition from the French government to carrying this out in the oil sector. To begin I briefly sketch the European position on energy markets, and then examine in more detail their position on the French oil sector.

In the 1970s the focus of European institutions was on tracking the poor performance of the refining sector in member countries because of the potential implications for regional employment and security of supply, rather than on the promotion of competition. On both questions it reached sanguine conclusions: regardless of the poor performance of the sector in the face of falling demand, underutilized capacity, and increasing petroleum product imports from non-member countries, intervention was unwarranted on employment or public security grounds, and "solutions ... will have to reflect the traditionally liberal attitude of the Community to trade policy."[108]

The Commission alluded to two potential justifications for public intervention in response to the poor performance of the sector. The first was to provide adjustment support for those made unemployed. Employment losses caused by the closing of thirty-five refineries were estimated by the Commission to be approximately 16,000 (although eighty percent of people losing their jobs were either reemployed or made redundant through early retirement). This meant total employment in the refining industry within member countries fell from 142,000 in 1975 to 126,000 in 1985, a reduction of eleven percent of total employment over the decade. The Commission did not, however, view this as significant enough to warrant recommending public intervention to the Council.[109] The second potential justification for policy

[108] Commission of the European Communities, "The Situation in the Oil-Refining Industry and the Impact of Petroleum Product Imports from Third Countries," COM (85) 32 final, March 1, 1985, 2.

[109] Commission of the European Communities, "The Situation in the Oil-Refining Industry," 3.

intervention lay in ensuring security of supply, but the Community was also sanguine about the implications of falling refinery capacity and rising imports. It found that the existing closing or mothballing of refineries, coupled with future projections of refinery closures, was unlikely to reach a level that would lead to concern over security of supply in final products derived from petroleum.

This assessment was confirmed at meetings of the European Energy Council on March 15, June 20, and November 11, 1985, at which the question of increasing imports of petroleum products from non-Community countries was taken up.[110] The decision by national governments to complete the internal market gave the Commission the opportunity to pursue the dismantling of trade and other regulations that acted as obstacles to trade in energy products and competition between firms across the energy sector. This implied the abolition, as the Commission asserted, not only of regulations hampering the free circulation of goods and services, but also implied harmonization of competition rules, abolition of subsidies provided to state monopolies, and of the laws and regulations governing such firms that impeded competition. The mandate acquired by the Commission covered the electricity and gas industries, as well as petroleum. At its request, the Council of the European Commission – which included the French Permanent Representative to Brussels Francois Scheer – accepted unanimously that the Commission should draw up a list of the obstacles that existed in national policy blocking the creation of an internal market for energy.[111]

The Commission opened its assessment of the national changes in legislation required to complete the internal energy market with a quotation taken directly from the 1968 guidelines:

> There are still considerable barriers to trade in energy products within the Community. If this state of affairs does not alter and if a common energy market is not achieved in the near future, the degree of integration achieved in this sector may well be jeopardized.[112]

110 Commission of the European Communities, "The Situation in the Oil-Refining Industry," 12.

111 Council of the European Communities, "1171st Meeting of the Council – Energy," June 2, 1987.

112 Council of the European Communities, "The Internal Energy Market," COM (88) 238 final, May 2, 1998, 2.

The decision of national governments to complete the Single European Market (SEM) gave the Community the mandate it required to recommend to the European Council that regulations and directives be issued in order to implement an internal energy market across member countries. At the Council of Energy Ministers meeting of June 2, 1987, the ministers of the national governments endorsed the Commission's request to identify obstacles to the completion of the SEM in energy, and then make recommendations to the Council on how to remove these in order to achieve the goal by the 1992 deadline. This mandate was noted in the Commission's subsequent report on the internal energy market:

> With the adoption of the Single Act in December 1985 and especially with the recent decision by the European Council to assign the resources needed for the Community to make a success of the Single Act, the way is at last clear for making a reality of the European internal market. Completion of the large market by the end of 1992 has become a key objective and the focal point of the revival of the European Community.[113]

The Commission pursued this mandate by issuing recommendations for the harmonization of policies that applied first to all member countries. Specifically, it stated that the creation of the SEM by 1992 required that regulations across product specifications for petroleum products that varied across the member states needed to be harmonized, and pointed to the problems that existed in the national regulatory schemes of individual member states. French national law featured prominently, and the Commission stated they should be removed for France to be in compliance with the intergovernmental agreement reached to form a single market for Europe by 1992.[114]

By the time the Commission made these recommendations, however, the major actors that supported and benefitted from the policies of national control were willing to accept European directives in petroleum. In fact neither firms nor policymakers saw benefit in maintaining the status quo. In the case of the firms, as noted above, the response to the oil shocks of the 1970s and subsequent collapse in demand within

[113] Ibid., 2.

[114] Commission of the European Communities, "The Community Oil Market, Its Oil Refining Industry, and the External Trade in Petroleum Products," COM (86) 263 final, May 13, 1986, 29.

France was to seek market share elsewhere and – in the case of Elf Aquitaine – to diversify out of the petroleum sector. State ownership of the firms proved a hindrance to these objectives, as did the restrictions on pricing and licensing structure that had given them market share. They therefore welcomed the jettisoning of strategic control, and at times lobbied for it to be moved forwards. For policymakers, the shift in the structure of the international oil market made the existing system of fixed prices untenable and led to the jettisoning of the fixed-price regime in favor of a floating price regime. The Directorate of Hydrocarbons, which had traditionally championed the interests of the national firms within the system of dirigisme, supported the relaxation of prices, in line with the position of Elf Aquitaine and Total. The Ministry of Finance also supported price liberalization as crude prices began to recede, believing this would help control inflation. There was no opposition, therefore, to these proposed changes by those who had previously benefitted from the institutions of national control. The head of hydrocarbon and natural gas issues in the Ministry of Industry argued:

> Europe had no influence whatsoever on the question of the system being dismantled. It was already being dismantled. Europe was always a train later. We were more rapid than them. The discussion with Brussels was on excluded sectors: telecom, electricity and so on, which fell under that regulation. So Europe had no influence whatsoever on energy regulation up to the changes that happened after 1986 with the excluded sectors in electricity and gas. On oil, they were not influencing it.[115]

Given this, the redrawing of the law of 1928 into the 1992 law governing the petroleum sector, and the privatization of the national-flag firms that was the center of the French strategy, proceeded with little opposition. Rather than attempting to retain articles in the legislation that maintained barriers to market entry designed to support French national firms, or resist directives from the European Commission, therefore, debate within the Directorate of Hydrocarbons on the new legislation designed to govern the oil sector focused on more mundane issues of ensuring compatibility with existing laws, and ensuring

[115] Gilles Bellec, interview with author.

consistent use of language.[116] Renegotiation of the existing policy led smoothly to the transformation of French oil market governance, rather than its adjustment as occurred in the case of Japan and the United States. This also meant that the idea of strategic control, as reflected in these policies, was jettisoned in favor of a new equilibrium focused on Europe as a response to the ongoing reliance on external markets for the supply of petroleum. For France oil was transformed into a commodity – certainly one with political implications, but nevertheless a commodity – rather than a strategic good requiring the intervention of the state.

European initiatives did have a more explicit effect on outcomes in two cases. First, the European Court of Justice (ECJ) ruled against the structure of the golden share (*action spécifique*) retained by the state in 1993 in preparation for privatization. The golden share gave the minister the authority to veto share acquisitions above a one-tenth, one-fifth, and one-third threshold, as well as halt asset sales. The European Commission notified the French government on May 15, 1998, that the golden share was incompatible to the ECJ, on the grounds that the state's special rights in relation to share acquisitions were not sufficiently precise.[117] It argued the golden share, as structured, was non-compliant with the free movement of capital (Article 56) and the right of establishment (Article 43) of the Treaty Establishing the European Community. The Minister of Economic Affairs, Finance and Industry responded on July 31 that the state would amend the structure of the golden share but that member states were able to ensure security of energy supplies through such measures in principle. A draft decree noted that the right of the minister could only be carried out when the acquisition "might threaten to disrupt France's supplies of petroleum products."[118] This was rejected as insufficient by the European Commission, however, and the matter was referred to the ECJ on July 28, 1999. In hearings, representatives of the European Commission argued that community law could be superseded by national law on the grounds of security of oil supplies, but that the means to achieve

[116] Senior policymaker in oil sector, interview with author, Paris, France, August 20, 2008.
[117] Commission of the European Communities vs. French Republic, Judgment of the Court of June 4, 2002. Case C-483/99.
[118] Ibid.

this end were disproportionate. The commission representative also argued that stockpiling measures in the International Energy Agency (IEA) and at the European Community level were adequate to manage supply risks. Representatives of the French government, on the other hand, argued that the structure of the golden share complied with European rules on the freedom of movement of capital. They further identified the importance of security of oil supplies, and the right of the state to guarantee those supplies, in arguing their case: "Petroleum products are of fundamental importance for a country's existence since not only its economy, but above all its institutions, its essential public services and even the survival of its inhabitants depend on them."[119]

The ECJ found on June 4, 2002, that ensuring security of oil supplies was a legitimate public interest, but that the golden share held by the state was disproportionate to this goal. The response of the Minister of Finance Francis Mer to the decision was sanguine, however: "the [merged firm of TotalFinaElf has] the means to take care of its own future. As a result, the issue is no longer dangerous for us."[120] TotalFinaElf did not officially comment on the decision when asked. Management did not view it as a significant change in their commercial position, rather viewing the share price and size of the firm as the best defense against being acquired. "It was the end of the privatization process, and it wasn't really perceived by the firms as a protection. The philosophy was rather that the best protection was the market capitalization of the firm."[121]

Alternative explanation – ideational change

A second alternative explanation focuses on the role of ideas. Policymakers clearly played an important role in transforming the laws and regulations used to govern the French oil market, and this is

119 Commission of the European Communities vs. French Republic, Judgment of the Court of June 4, 2002. Case C-483/99, I-4799.

120 Stephen Castle, "European Court Outlaws France's Golden Share in Elf," *The Independent* (June 5, 2002): 20. See also Adrien De Tricornot, "M. Mer accueille sans inquiétude la remise en cause de la Golden Share de l'État dans TotalFinaElf," *Le Monde* (June 6, 2002): 21.

121 Jean du Rusquec, interview with author; Xavier Preel, General Manager, Total Exploration and Production, interview with author, Paris, France, July 29, 2008.

correlated with the shift in their position towards the governance of economic markets in general terms. Most obviously, the broader redefinition of the relationship between government and firms determined the timing of the sale of equity in Elf Aquitaine and Total shares in the case of privatization, just as this shift played a role in the privatization of oil assets in Japan described in the next chapter. If we focus directly on the oil industry, there is also some evidence that policymakers recognized that the changes in the international oil market made protecting domestic market share, and expanding CFP/Total and Elf Aquitaine shares of oil produced internationally, no longer a rational response to the management of risk in the international oil market. The councilor advising the State Secretary for Energy, for example, noted that he supported price liberalization because it made little sense given the transformation of the oil market.[122] Yet the evidence above suggests that the distributive effects of oil price volatility and its implications for economic growth had a profound effect on the strategy adopted by policymakers. The decision to relax prices, for example, was agreed on and implemented in order to take advantage of falling crude oil prices. The decision to relax licensing was also partly motivated by the desire to allow new market entrants – in the form of supermarkets – in order to increase competition and lower gasoline prices.[123]

Firms mattered in two ways. First, where existing laws and policies hindered rather than supported firm profitability, the firms actively promoted liberalization. Most notably, the relaxation of fixed prices on oil products in the early to mid-1980s occurred as a direct result of lobbying by firms, which suffered commercially because the prices of inputs were made volatile by nationalization, while the prices of outputs continued to be fixed. Second, firms invested no resources in retaining state ownership or licensing regulations, and indeed supported both measures. Privatization was judged by management to be in the interest of the firms given that the infusion of capital by the state through equity ownership was a burden rather than a benefit to management as it expanded its operations internationally. The regulation of domestic market share, on the other hand, became unimportant given the diversified structure of business and the poor performance of the refining sector within France. Instead, both CFP/Total and Elf

[122] Gilles Darmois, interview with author.
[123] Jean du Rusquec, interview with author.

Aquitaine pursued a strategy that focused on expanding and integrating operations within Europe and internationally, rather than defending market share within the French domestic market from import competition.

In the next chapter I move on to examine the politics of oil market governance in Japan. The Japanese case offers useful comparative evidence in support of the argument that firm characteristics are shaped by negotiations between governments and firms negotiated in earlier periods. The responses of most Japanese firms to the transformation of the international oil market also offer an important contrast to the French case: the smaller and less diversified structure of Japanese firms made the types of demands they made on governments significantly different to those of French firms. While they did not win everything they sought from government, these demands meant adjustments were made to the industrial compact used to govern oil markets in Japan, but these adjustments did not amount to the abandonment of attempts to improve the competitiveness of firms.

5 *Adjusting oil market governance in Japan*

A tail wind blows only for those who know its direction.
[Kaze no fuku hoko wo shiru mono ni nomi, oikaze wa fuku]

Liberal Democratic Party report on energy security, November 1978

In this chapter I identify and explain changes in oil market governance in Japan. Japan is a useful case through which to compare the effects of differences in firm characteristics on forms of oil market governance. Japan, like France, almost wholly lacks domestic reserves of petroleum. In Japan, as in France, policymakers sought to improve the position of domestic firms relative to the affiliates of the majors. Indeed, initial laws designed to redistribute market share were informed by the 1928 French oil law.[1] Unlike France, however, the liberalization of Japan's oil markets was partial; the domestic market remained open following the liberalization of trade in oil and oil products implemented between 1986 and 1996, although subsidies in support of national firms' exploration and production internationally were reorganized but remained in place. Policies also continued to promote the vertical integration of Japanese firms engaged in the oil sector, with the goal of creating a Japanese oil major.

What explains the difference between outcomes in France and Japan, and what explains the adjustments made to the industrial compact negotiated between business and government? In this chapter I argue that ongoing intervention in the oil sector in Japan is shaped by the characteristics of Japanese firms, and I discuss the way this affects the demands they make on government. I also argue that the characteristics of Japanese firms were shaped by the industrial compact negotiated between business and government prior to the shift in

1 Takeo Kikkawa, "1934-nen no Nihon no Sekiyugyoho to Sutandado Bakyumu Oiru Kampani (1) [The 1934 Petroleum Industry Law and Standard Vacuum Oil Company (1)]," *Aoyama Keiei Ronshu* 23 (1989): 21–43.

the structure of the international oil market. Firms operating in exploration were small and poorly diversified, and continued to support subsidies in order to improve their competitive position. Firms operating in refining and marketing were also small and poorly diversified, with few assets in exploration and production. When the problem of how to adjust the laws and policies governing Japan's oil market was raised, therefore, almost all firms sought to retain government support in order to improve their competitive position.

Firms did not get everything they wanted during the debate over how to reconfigure oil policies in response to the shift in oil market structure. While domestic refiners did not face the problem of nationalization that hurt French firms, price volatility drove changes in demand that undermined their performance, as was the case in both Europe and the United States. Unlike French firms, however, Japan's refiners were not in a position to make up for losses at home by diversifying into other markets. As a result they continued to demand that the government protect them from competition. Industry crisis, and counter-demands from manufacturers to lower the price of inputs, reduced the willingness of policymakers to maintain barriers to trade in oil and oil products. Policymakers' willingness to retain the industrial compact negotiated between them and firms unchanged was also weakened by the failure of the existing laws governing the domestic market to improve the international competitiveness of those firms. As a result, support grew for exposing rather than shielding these firms from competition. This, it was hoped, would force mergers and make domestic firms more competitive.

The renegotiation of Japan's industrial compact governing the domestic market matched changes made to the policies used to support the activities of Japanese firms operating in exploration and production. Adjustments were made to the structure of subsidy support, but they did not amount to a rejection of promoting national firms. Partial privatization was not primarily driven by ideational change, nor by firms rejecting policy support because of increased firm size and diversification. Rather, it was driven by a restructuring of the policy instruments used to support firms because of their poor performance. Given firms' ongoing lack of competitiveness and continued demand for support from the state, this left the door open to the reinvigoration of subsidy support for these firms in the 2000s.

What of the origins of firm characteristics? In the previous chapter I argued that the organization of French firms was substantially influenced by the industrial compact negotiated to regulate market share, and that this went on to condition the response of those firms to the shift in the structure of the international oil market. In the first section of this chapter I similarly investigate what led to the comparatively small size of Japanese firms and their lack of diversification. I argue that, as in the case of France, the characteristics of Japanese firms was substantially influenced by the laws and policies negotiated between business and government. One simple way to investigate this is by examining what happened to refining firms following the abolition of the most important law governing the domestic market. Evidence presented later in the chapter shows there were substantial mergers, and other forms of corporate tie-ups, following the relaxation of restrictions on oil product imports, suggesting import barriers were having an important effect on firm characteristics. As I describe below, support provided to firms was also provided on a non-discriminatory basis and, in the case of subsidies used to reduce the risks of international exploration and production. This encouraged the fragmentation of firms rather than promoting the incorporation of multiple projects within a single firm in order to diversify risk. The result was a horizontally and vertically fragmented industry that was in a poor position to take advantage of the new opportunities presented by the shift in oil market structure, and to manage its risks.

Oil market governance in Japan

As in France, negotiations over how to govern the oil market were prompted by the commercial position of the affiliates of the majors within the domestic market. One response to the competitive threat presented was cartelization. A second focused on protecting the market share of domestic refiners and marketers through regulation. Critically, the industrial compact negotiated between firms and the government tended to cement vertical and horizontal fragmentation, in contrast to France.

The consumption of oil products in Japan shifted from kerosene to light oils as their use increased in the fishing industry It also increased demand for gasoline and heavy oil.[2] As in France, demand growth was not matched by domestic oil production, which peaked at eighty-three percent of consumption following the introduction of rotary drilling in 1918.[3] The increase in demand was therefore met largely by imports, which rose to fifty-six percent of total demand by 1926, from thirty-seven percent in 1924.[4]

In terms of the organization of supply, foreign affiliates of the majors held a dominant position in the domestic market, in common with the French case, and imported the majority of their product rather than investing in refining infrastructure within Japan.[5] In the four-member kerosene cartel established in 1910, for example, the share of domestic sales enjoyed by Standard Oil (New York) was forty-three percent, while Royal Dutch Shell affiliate Rising Sun was allocated twenty-two percent. Two domestic firms were allocated thirty-five percent of the market. In the six-member gasoline cartel established in 1932, on the other hand, Rising Sun enjoyed the largest market share with thirty-two percent, with Socony Vacuum allocated with twenty-one percent of the market.[6] The largest domestic supplier

[2] Gasoline and heavy oil consumption increased from 555,000 kiloliters in 1930 to over one million kiloliters in 1938. Over the same period heavy oil consumption increased from 1.1 billion to 2.6 billion liters. Tosuke Iguchi, *Nihon Sangyo Hattatsushi: Sekiyu [The History of Industrial Development in Japan: Petroleum]* (Tokyo: Gendai Nihon Sangyo Hattatsushi Kenkyukai, 1963), 245.

[3] Nihon Sekiyu, *Nihon Sekiyu Hyakunenshi [A One Hundred Year History of Petroleum]* (Tokyo: Nihon Sekiyu Kabushikigaisha, 1988), 208. For an English-language review of this period that focuses on firm–state relations see Richard J. Samuels, *The Business of the Japanese State: Energy Markets in Comparative and Historical Perspective* (Ithaca: Cornell University Press, 1987); Yul Sohn, *Japanese Industrial Governance: Protectionism and the Licensing State* (London: RoutledgeCurzon, 2005).

[4] Nihon Sekiyu, *Nihon Sekiyu Hyakunenshi*, 285–6.

[5] Domestic firms began to invest in refining capacity in the 1890s. Takeo Kikkawa, *Nihon Sekiyu Sangyo no Kyosoryoku Kochiku [Building the Competitiveness of Japan's Oil Industry]* (Tokyo: Nagoya Daigaku Shuppankai, 2012), 17.

[6] Allocations shifted to the benefit of domestic refiners in later agreements. The cartel expanded to include Matsukawa Oil in 1934, which had contributed to falling prices by importing gasoline from the Soviet Union. See Takeo Kikkawa, "1934-nen no Nihon no Sekiyugyoho to Sutandado Bakyumu Oiru Kampani (2) [The 1934 Petroleum Industry Law and Standard Vacuum Oil Company (2)]," *Aoyama Keiei Ronshu* 24 (1989): 77; Takeo Kikkawa, "1934-nen no Nihon no Sekiyugyoho to Sutandado Bakyumu Oiru Kampani (3) [The 1934

was Nihon Sekiyu, which held the second largest market share with twenty-four percent.[7]

Government interest in national oil market regulation emerged in the Department of the Navy, which proposed in 1918 that the domestic oil market should be regulated and that domestic oil firms should be merged.[8] In 1926 the incentive to refine domestically was increased by cutting duties on crude imports and raising them on products. In the same year Cabinet established a national fuels committee within the Ministry of Commerce and Industry (MCI). It reported to the minister in 1928 that the market shares of Rising Sun and Socony represented a threat, and emphasized the importance of improving the competitiveness of domestic industry by merging the assets of importers, refiners, and distributors – including foreign affiliates' – into a single entity.[9] The minister ordered further study of the most appropriate national fuels policy, which recommended that the government regulate imports of crude and petroleum products through licensing, restricting entry into the domestic refining market, subsidizing exploration and production domestically, promoting firms internationally, and subsidizing the production of substitutes such as ethanol and oil shale.[10]

Negotiations over how to regulate the domestic market began in earnest in the 1930s. In 1933 the government released two proposals for governing the oil industry. The first almost perfectly mirrored France, proposing a national monopoly of the production, trade, and sale of oil and oil products, with rights delegated to a national-flag company with public–private participation. Under the proposal, the assets of foreign firms would be forcibly acquired by the government.

Petroleum Industry Law and Standard Vacuum Oil Company (3)]," *Aoyama Keiei Ronshu* 24 (1989): 40–1. In October 1933 Nihon Sekiyu, Ogura, and Mitsubishi Sekiyu also proposed merging in response to their difficult commercial position. See Staff reporter, "San Taikoku Sekiyu Gaisha, Godo ni Iken Icchi: Shorai ha Chusho Kaisha no Daigodo he [Three Petroleum Firms Agree to Merge: The Creation of a Central Firm in the Future," *Asahi Shinbun* (October 24, 1933): 4.

7 Figures include Taiwan and Korea. Kikkawa, "1934-nen no Nihon no Sekiyugyoho to Sutandado Bakyumu Oiru Kampani (1)," 23.

8 Ibid., 29–30.

9 Ibid., 35. Kikkawa argues this shows the interests of domestic firms did not dominate the outcome, as stripping refining from domestic firms harmed the interests of Nihon Sekiyu – the industry representative on the committee – which had just invested in a new plant.

10 Nihon Sekiyu, *Nihon Sekiyu Hyakunenshi*, 304.

The law followed the second of the proposals put forward by the MCI, however, rejecting a state monopoly and instead establishing a system of licensing for the right to import and construct refinery capacity. Customs duties remained higher for crude than for oil products in order to encourage further development of the domestic refining industry.

Industry sought cartelization and resisted mergers, primarily as a response to falling product prices. In 1932 an agreement was struck with the support of MCI between six importers and domestic refiners (Socony, Rising Sun, Nihon Sekiyu, Ogura Sekiyu, Mitsubishi Sekiyu, and Mitsui Bussan) to allocate gasoline sales, as noted above.[11] Management in both Socony and Rising Sun sought to enlist the support of governments in Great Britain, Holland, and the United States in protest against the requirement that market participants maintain six months inventory in Japan as a physical hedge against supply disruptions. Evidence suggests the efforts of the majors, backed by the US government, to influence the implementation of the law met with some success.[12] Diplomatic efforts by the US government exempted the firms from this requirement, but the firms also feared they would be discriminated against in the distribution of market share. This occurred through their loss of a share of demand growth in the first allocation of quotas. Later allocations kept the market shares of foreign firms constant, distributing incremental growth to domestic firms with refining capacity and storage facilities in Japan.[13]

[11] Cartelization in oil matched the industrial strategy adopted across a wide range of sectors in the prewar period. For a summary see Takeo Kikkawa, "Senzen Nhon no Karuteru [Pre-war Japanese Cartels]," *Aoyama Keiei Ronshu* 25 (1991): 73–98.

[12] A 1945 Senate review of US diplomatic policy summarized the outcome as follows: "The American Embassy at Tokyo, on behalf of the American company, protested vigorously and repeatedly to the Japanese government over a period of some 3 years, as did the British Embassy ... We did not succeed in inducing the Japanese Government to abrogate or modify the law (the military control over the civilian government was too strong for that) but we did succeed in inducing that government to postpone indefinitely the application to the two foreign companies of the stockholding provision of the law." See "American Petroleum Interests in Foreign Countries," on June 27–28, to the Hearings before a Special Committee Investigating Petroleum Resources, US Senate 79th Congress, 1st session, 1945, 320.

[13] Irvine H. Anderson, Jr., *The Standard-Vacuum Oil Company and United States East Asian Policy 1933–1941* (Princeton University Press, 1975), 102 fn. 86. Socony and Rising Sun refused to participate in the gasoline cartel arrangements in 1934 in protest. Both the firms, and the US government,

In contrast to the move towards redistributing market share among existing firms within Japan proper, a monopoly model was adopted in Manchuria and the Korean Peninsula. In Manchuria, for example, the Manchuria Petroleum Company was established in 1934 with shareholding by the South Manchuria Railway and domestic oil firms, and headed by the president of Nihon Sekiyu. It was given a monopoly over oil production, refining, and distribution within Japan's area of effective control. The US government opposed the arrangement, arguing that it hurt US commercial interests and contravened Article 3 of the Nine Power Treaty of 1922, establishing the principle of commercial equality throughout China.[14]

A second component of the industrial compact negotiated in Japan, as in the case of France, focused on exploring for and producing oil. The Department of the Navy promoted the organization of private oil interests in Sakhalin in 1919 through the Hokushinkai, with 200 employees and the participation of a number of private firms.[15] Its assets were eventually transferred to the North Karafuto Oil Company. Exploration efforts met with some success: production increased from 33,000 tons in 1926 to a peak of 193,000 tons in 1933.[16] Yet there was no coordination of the Sakhalin development with the strategies adopted on the Korean Peninsula or in Manchuria.

The 1934 oil law, in addition to regulating the domestic market in Japan proper, required the development of an upstream strategy. A 1937 proposal for upstream development included subsidies for firms operating both domestically and internationally. Subsides were provided from 1937 for firms engaged in exploration and production, and the Sakhalin development was allocated almost thirteen million

pushed over 1934–5 to have their market shares guaranteed under the new scheme, but were unsuccessful. Kikkawa, "1934-nen no Nihon no Sekiyugyoho to Sutandado Bakyumu Oiru Kampani (3)," 60–1.

14 "Informal Memorandum," American Embassy in Japan to the Japanese Ministry of Foreign Affairs, 893.6363 Manchuria 29, in *FRUS Japan: 1931–1941 Volume I*, 130; "Oral Statements by the American Ambassador in Japan (Grew) to the Japanese Minister for Foreign Affairs (Hirota)," 893.6363 Manchuria/194, in *FRUS Japan: 1931–1941 Volume I*, 150–1. On Korea see Nihon Sekiyu, *Nihon Sekiyu Hyakunenshi*, 315–16.

15 Administrative responsibility for the project was transferred to MCI in 1926.

16 Tomio Noda, "Nenryo Kokusaku to Sekiyu Shigen Kaihatsu: Kita Karafuto Sekiyu Kabushikigaisha to Teikoku Sekiyu Kabushikigaisha [National Fuels Policy and the Development of Oil Resources: The Kita Karufuto Oil Company and Imperial Oil Company]," *Journal of Political Economy* 70 (2004): 53–72.

yen over a five-year period from 1937.[17] On March 28, 1938, the Petroleum Resources Development Law institutionalized a planning process for upstream development, requiring firms to submit business plans, and offering subsidies up to two-thirds of costs. The 1934 and 1938 laws did not seek to force firm mergers, however, and the government did not take a stake in firms engaged in upstream development. The most important effort to coordinate investments and spread risk in exploration and production activities was the Imperial Oil Development Corporation. The firm was established in 1940, and had Nihon Sekiyu with forty-five percent equity, Nihon Kogyo with thirty percent, and other firms taking a five percent stake. It was also small, however, with just 130 employees at its outset.[18]

The onset of the Pacific War and ensuing occupation reordered governance in the oil sector. In the downstream sector, the outbreak of war achieved in the short term what the first law was only partially successful in doing: increasing the control of national firms over the domestic market for petroleum products. Where the prewar institutions promoted cartelization, and thus ongoing fragmentation, the domestic refining industry was converted into a monopoly in September 1939. This structure continued for nine years before being disbanded by the occupying authorities in March 1949.

Increased control over refining and marketing was useless, however, without adequate supplies of crude. Here firms and policymakers faced the problem identified some twenty years earlier by military planners as Japan's Achilles' heel. Until World War I, non-belligerents typically provided the physical and financial resources required to wage war. That this was no longer the case was brought home starkly to the Japanese government by the September 1941 decision by the United States to embargo the export of oil to Japan. By the end of the war Imperial Oil controlled ninety-one percent of Japan's meager domestic production, but efforts to secure resources internationally floundered. Domestic refiners were left with much of their capacity destroyed, and demand was anemic.[19] In January 1946 the Supreme

[17] Project development eventually ceased as Soviet–Japanese relations worsened.

[18] Noda, "Nenryo Kokusaku to Sekiyu Shigen Kaihatsu."

[19] Staff reporter, *Oil & Gas Journal*, December 26, 1955, 173–4. Cited in Chih Chen Ching, "Crude Oil Prices and the Postwar Japanese Refining Industry" (PhD diss., Massachusetts Institute of Technology, 1967), 165.

Commander of Allied Powers (SCAP) banned crude oil imports through SCAPIN 640, and in October the same year SCAPIN 1236 ordered the closing of refinery capacity that remained operational on the Pacific Coast.

Opinion was divided within SCAP on whether oil infrastructure should be included in reparations, but the decision was taken in December 1946 to leave it intact.[20] The most significant change implemented during the occupation was the completion of a series of tie-ups in refining between Japanese and US firms. Socony and Rising Sun imported and marketed products, or sold crude, to Japanese marketing firms in the prewar period. In the years during and after the occupation, foreign firms entered into direct agreements with Japanese firms. Evidence suggests these joint ventures made sense to domestic firms and the majors, and were not forced on the firms or on the Japanese government by SCAP.[21] Rather, foreign firms had access to the crude that Japanese refiners needed to meet demand, and domestic firms were too weak to finance the expansion of refinery capacity. Between 1949 and 1952, domestic firms partnered with foreign firms, most notably Nihon Sekiyu, which partnered with Caltex, and Toa Fuels, which partnered with Standard Vacuum. These cooperative relationships took the form of joint ventures, as required by the Law Concerning Foreign Investment. Refining and marketing firms, such as Idemitsu Kosan, that did not enter into tie-ups with foreign affiliates gained access to capital through loans from the majors, or from banks with the majors acting as cosignatories, in return for guarantees of the purchase of crude.[22]

[20] Yoshida argues the US majors in the Petroleum Advisory Group (PAG) to SCAP sought to harm domestic refiners by recommending their facilities be scrapped. Kikkawa rejects this view. Regardless, the recommendation to scrap refining facilities included in the second Strike Report had no influence on occupation oil policies. See Takeo Kikkawa, "GHQ no Sekiyu Seisaku to Mejazu no Tainichi Senryaku [GHQ's Oil Policy and the Majors' Strategy Towards Japan]," *Aoyama Keiei Ronshu* 26 (1991): 265–84.

[21] Ibid., 277.

[22] The Law Concerning Foreign Investment was passed in May 1950 with the goal of encouraging investment in capital-poor Japan while maintaining control over foreign currency reserves. The law stated that investment was allowed only if it contributed to the self-support and sound development of the Japanese economy as well as to the improvement of the international balance of payments. On the logic of the tie-ups see Ching, "Crude Oil Prices and the Postwar Japanese Refining Industry," 182–4.

Postwar oil market governance

As in France, postwar oil policy reconstituted the approach taken in the prewar period. It was also, like in France, embedded in a broader effort to reindustrialize. An important part of the postwar growth strategy was the avoidance of "excess competition" through limiting market entry and avoiding bankruptcies.[23] MCI was reconstituted during the occupation as the Ministry of International Trade and Industry (MITI), with the mission of using industrial targeting to promote the competitiveness of firms across manufacturing, as well as in the energy sector.[24] Industrial targeting focused initially on increasing output in coal and steel in order to aid economic recovery. The coverage of industrial policy expanded over time to include synthetic fibers, petrochemicals, autos, shipbuilding, machine parts, and electronics.[25] A standard set of policy instruments was used, focusing on quotas and tariffs, subsidies, and the relaxation of competition rules.

The energy sector was no different. The compact negotiated between business and government remained similar in the postwar period, despite the entrance of affiliates of the majors into refining through tie-ups with domestic firms. Trade barriers and subsidies were reestablished in order to protect and nurture domestic firms. In 1961, Cabinet announced it would liberalize the importation of crude oil, but also introduce a new industry law to govern the domestic market. The law gave the state the authority to influence supply and pricing decisions. It mimicked the prewar law, establishing a licensing system for crude and crude product imports, and the distribution of these goods within the Japanese market.[26]

23 Ryutaro Komiya, "Introduction," in *Industrial Policy of Japan*, eds. Ryutaro Komiya, Masahiro Okuno, and Kotaro Suzumura (Tokyo: Academic Press Japan, Inc., 1988), 11–13.

24 I make no claim about the efficacy of this strategy in promoting growth. In fact, as I detail below, the effectiveness of government policy in the oil sector was poor. More generally, whether Japan's model of bureaucratic delegation and sectoral targeting was responsible for the extraordinary growth the country experienced remains open to debate. For a general review of industrial policy see Komiya, Okuno, and Suzumura, (eds.), *Industrial Policy of Japan*.

25 Yutaka Kosai, "The Reconstruction Period," in *Industrial Policy of Japan*, Komiya, Okuno, and Suzumura (eds.), 25–43.

26 The bill was passed by the Cabinet on March 6, 1962. Compromise with opponents led a clause initially to be attached stating that the law could be changed or abolished depending on changes in external market conditions. Industry lost control of the debate in parliament, however, and the clause

In industry hearings, the majority of firms agreed with the introduction of the law. Among refiners and marketers, seventeen domestic firms agreed with the law in whole or in part, while Idemitsu joined with Caltex and Shell, and was the only domestically focused firm in opposition.[27] The central feature of the new law was a five-year rolling plan that calculated the production and imports of crude oil and oil products, and refining capacity. Refining firms were required by law to lodge plans with the regulator, which could order changes to these plans. The law also conferred the power to regulate the construction of new refining capacity, and mergers and acquisitions. Although imports of petroleum products were allowed, they were intended solely to meet gaps between planned domestic production and demand. This occurred most commonly in the cases of naphtha and residual fuel oils.[28]

Together, this amounted to the archetype of plan-rational *étatisme*. The competent minister was conferred the authority to order firms to adjust import plans in order to solve mismatches between planned consumption and production. In practice, the industry arranged adjustments through the industry association, with government officials coordinating if the firms were unable to reach agreement among themselves.[29] One effect of the law was to shift the basis of competition between firms from price to refining capacity. Production schedules were determined using a formula that included installed refining capacity, meaning that constructing refining capacity guaranteed an increase in market share. This made competition for licenses strong: in the first year following the passing of the law applications were made to increase refining capacity by more than one million barrels per day,

was dropped during the committee phase. This meant that the law came into force on July 10 with no implicit sunset clause. Nihon Sekiyu, *Nihon Sekiyu Hyakunenshi*, 591–5.

[27] Akira Yamaoka, "Sekiyu Gyoho Kara Kyohan Gyosha Koso he [From the Petroleum Industry Law to the Joint Sales Company]," *Keizai Ronso* 170 (2002), 50–1. Yamaoka claims that industry division enabled the government to shape the law towards its favored outcome.

[28] Tsutomu Toichi, *OPEC Export Refineries and the Implications for Japan's Petroleum Policy* (Cambridge, MA: MIT Press, 1985).

[29] Negotiations were not always smooth. In 1964, for example, MITI threatened to order Idemitsu Kosan to adjust production after the latter left the industry association in protest against production orders. It was not required to use its powers after Idemitsu Kosan relented.

of which a little over 420,000 barrels were approved. Between 1963 and 1973 the applications-to-approvals ratio fluctuated between fifteen and forty percent, offering further evidence of the importance of licenses.[30] The system also ensured that the market share of independent Japanese firms, when measured in terms of refining capacity, increased relative to firms with tie-ups with the non-Japanese majors, from approximately forty to fifty percent between 1962 and 1973.

A second component of the industrial compact focused on promoting mergers in order to develop a market leader, although this was less successful than in the case of France. A step towards this goal was achieved in 1965, when Kyodo Petroleum was established from national firms Asia Petroleum, Japan Mining, and Toa Petroleum.[31] The firms merged distribution and sales only, however, and together they enjoyed only a small share of the market across oil products, with six percent of national storage capacity, and 9.5 percent of sales.[32] The regulator sought to increase the competitiveness of the firm by using the 1962 industry law to preferentially allocate refining capacity to them, taking market share away from both domestic and foreign firms.[33] Yet this strategy harmed rather than helped firm performance because Kyodo was unable to increase sales, meaning growing refining capacity led to lower capacity utilization. Despite robust demand growth, this led to an imbalance between firms with strong sales but inadequate refining capacity, and those with weaker sales but more refining.[34]

Efforts to promote national-flag firms were not limited to refining and distribution. Domestically, Imperial Oil and JAPEX developed Japan's limited reserves. Internationally, a goal was established that thirty percent of domestic demand should be owned by domestic firms.

[30] Nihon Sekiyu, *Nihon Sekiyu Hyakunenshi*, 641. This led to conflict between firms and the ministry over the setting of aggregate capacity, and the allocation of capacity between firms. See, for example, Staff reporter, "Sekiyu Seisei Gyogai: Tsusanshoan Nomu [Refining Industry Accepts MITI Plan]," *Asahi Shinbun*, May 5, 1963: 4.

[31] Staff reporter, "Meisho ha 'Kyodo Sekiyu' [The Name is 'Kyodo Sekiyu']," *Asahi Shinbun*, April 18, 1965, 7. A few small firms later also joined.

[32] Akira Yamaoka, "Kyodo Sekiyu Gurupu no Tenkai [The Development of the Kyodo Oil Group]," *Keizai Ronso* 171 (2003): 42–3.

[33] See, for example, Nihon Sekiyu, *Nihon Sekiyu Hyakunenshi*, 642 and 650.

[34] Akira Yamaoka, "Kyodo Sekiyu Gurupu no Tenkai," 59; Nihon Sekiyu, *Nihon Sekiyu Hyakunenshi*, 642.

This was used as justification for establishing subsidies and discounted loans for firms operating in exploration and production. In 1967 the Japan Petroleum Development Corporation (JPDC) was created to provide cheap financing for exploratory activities. The firm did not operate upstream projects directly, however, and also did not discriminate between firms in order to promote mergers. As a result, by 1971 Japanese firms were involved in twenty-one petroleum projects globally across the exploration, development, and production phases, and by 1976 this increased to more than sixty projects.[35] Yet these were split among twenty firms, many of which were established to manage the specific project under development, with an average paid-in capital of 4.3 billion yen, or forty-three million dollars at one hundred yen to the dollar.[36] Unlike in France there was also no provision for domestic refiners to buy oil produced by Japanese firms upstream.[37]

By the late 1960s, therefore, the full panoply of policy instruments was applied in support of domestic firms. Government support helped establish a national presence in the exploration and production of oil internationally. Incentives to shift upstream were provided through a national oil company with upstream presence financially, but which did not act as an operator. Trade instruments restricted imports and exports in ways that strengthened the domestic refining industry, and regulatory instruments gave the government the authority to regulate sectoral investments and prices.

Yet the industrial compact negotiated in Japan met with mixed success. Support was provided to firms indiscriminately. While this helped

[35] Shigen Kenkyu Iinkai, "Hyo 8–3: Wagakuni no Chiiki Betsu, Shigen Betsu, Kaigai Kaihatsu Projekuto [Graph 8–3: Japan's Overseas Development Projects by Region and Resource-type]," in *Shigen Kenkyu Iinkai Kyoryoku Kaihatsu Mondai Senmon Iinkai: Shingi Ronten [Natural Resource Research Committee Cooperative Development Problems Specialist Committee: Debate Summary]* (Tokyo: Ministry of International Trade and Industry, November 24, 1971); Sekiyu Kogyo Renmei Kikaku Chosabu, *Wagakuni Sekiyu Kaihatsu no Genjo [The State of Japan's Petroleum Development]* (Tokyo: Sekiyu Kogyo Renmei, 1976).

[36] Shigen Kenkyu Iinkai, "Hyo A-3: Kaigai Sekiyu Kaihatsu Jigyo Ichiran [Graph A-3: Overseas Petroleum Development Projects]," in *Shigen Kenkyu Iinkai Kyoryoku Kaihatsu Mondai Senmon Iinkai: Shingi Ronten [Natural Resource Research Committee Cooperative Development Problems Specialist Committee: Debate Summary]* (Tokyo: Ministry of International Trade and Industry, November 24, 1971).

[37] Nihon Sekiyu, *Nihon Sekiyu Hyakunenshi*, 642 and 652.

firms to operate internationally in upstream exploration and production, it did not promote integration between the upstream and downstream. It also did not promote mergers between firms. In other words, it increased the amount of oil upstream owned by Japanese firms either as equity partner or as operator, but did not lead to the creation of a vertically integrated Japanese national oil company at scale; Japanese firms' operations remained fragmented across projects.

The problem of horizontal fragmentation extended to refining and distribution. Industrial strategy increased the share of the market controlled by domestic refining and distribution interests. However, as in the upstream, this was not done in order to promote the interests of a small number of firms, or a single national oil champion. As a result the refining industry was fragmented, and there was no significant vertically integrated firm; indigenous firms (those without tie-ups with foreign refiners) held 44.4 percent of total domestic sales by 1970, and owned fifty percent of total refining capacity, but there were twenty separate firms in total among the domestic firms, organized through cross-shareholding into ten groups, and twenty firms with foreign tie-ups, organized across four groups.[38]

The structure of firms operating in the oil sector in the Japanese case thus differed from the French case, despite similar initial conditions. What explains this difference? First it is worth noting that it cannot be attributed to the existence of French colonial possessions in North Africa with reserves of oil, or CFP's participation in the Middle East, for the simple reason that Japanese firms also secured shares in resources internationally. The horizontal and vertical fragmentation of the Japanese industry can also not be attributed to policies implemented following defeat in war, given that it preceded Japan's occupation. Instead, I have suggested that laws negotiated between firms and used to regulate domestic market share, and promote the interests of Japanese firms internationally, tended to consolidate the small size of firms, and their lack of diversification. Changes in the size and number of firms following changes to these laws in the 1980s and 1990s provides further evidence in support of this contention, and are introduced below. Below I examine below how the characteristics of Japanese firms went on to influence their responses to

[38] Idemitsu Kosan, "J-31: Wagakuni Sekiyugaisha no Gaiyo [J-31: Outline of Japan's Petroleum Companies]," in *Idemitsu Sekiyu Shiryo Showa 51-nenban [Idemitsu Petroleum Materials 1976]* (Tokyo: Idemitsu, 1976), 123–6.

the transformation of the international oil market, including the strategy they adopted towards government.

Adjusting oil market governance in Japan

The questions this section takes up are: why did the institutions of oil market governance continue to support intervention in the oil market in favor of national firms, and why were adjustments made to them, given the ongoing support of firms and policymakers? From 1980 the government adjusted, but did not abolish, support for domestic industry. This represented an important break from the past. Other than the interlude of the occupation, policy either maintained or increased barriers to market entry by non-domestic firms. In contrast, support for firms was cut in the 1980s, and by 1996 instruments giving the government authority to regulate trade, and entry and exit into the domestic refining sector, were all but abolished. Significant mergers took place as a result.

Government support for exploration and production by Japanese firms was also restructured, but was not abolished. The major state firm supporting private sector firms' efforts was reorganized, and a number of the firms that had been created and supported through its financing were privatized. Taken together, these changes represent a partial convergence on liberal forms of economic governance. Yet, despite this, policymakers continued to promote firms through subsidies and through direct ownership, and reinvigorated support for firms as oil prices increased in the 2000s.

In this section I show that the small size of Japanese firms, and their lack of diversification, meant they were ill-equipped to take advantage of new opportunities internationally, and had few options in the face of oil price volatility in the 1970s and 1980s. As a result, a common firm response to changes in the international oil market – rather than diversify internationally independently from government support – was to continue demanding support from policymakers. I also show that the willingness of policymakers to supply this support without concessions from firms fell. As in France, the rejection of industrial policy as a method for promoting economic growth was one reason for the change in policymaker preferences. Yet I find policymakers shifted their preferred policies because the transformation of the oil market undermined the performance of firms operating in refining and marketing, and internationally, thereby demonstrating that the existing strategy was failing. The weakness of firms, and their inability to respond to the change in oil market structure, therefore led policymakers to adjust, but not abandon, government support for domestic firms.

Markets and firm strategy

The transformation of the international oil market was synonymous with the diversification of oil production, nationalization, and price volatility. Evidence shows the small size and lack of diversification of Japanese firms meant corporate strategies differed from those adopted by French firms. Domestic firms operating in refining and distribution made few attempts to expand into oil production; when they did so it was in the form of equity rather than as operator. This was the case both for firms in tie-ups with the majors and purely domestic firms, suggesting this reluctance was not simply an artifact of the domestic refiners' relationship with the majors but was driven by other factors; the operational skills required in oil exploration and development are very different from those of refining and distribution, for example, and the performance of refiners and marketers was quite poor relative to manufacturing sectors throughout the 1960s, despite growing product demand and falling crude prices.[39]

Increased opportunities in the 1960s led some of these firms to increase investments internationally. Japanese firms participated in consortia investing in Malaysia, Indonesia, Thailand, Canada, Alaska, Abu Dhabi, Qatar, Egypt, and elsewhere. A total of nine firms were established during the 1960s, although most were single-project companies. The number of firms investing internationally increased markedly following the introduction of state subsides: a further forty-six firms were established to invest upstream between 1970 and 1976, and thirty-one of these were created between 1970 and October 1973 when the first oil crisis led to a spike in prices.[40] Nihon Sekiyu, for example, engaged in exploratory operations in Texas, California, and other states, planning an investment of about 300 million dollars, although it was not operator, with the stated goal of increasing

[39] Idemitsu Kosan, *Idemitsu Sekiyu Shiryo Showa 51-nenban*, 111.

[40] Sekiyu Kogyo Renmei, *Wagakuni Sekiyu Kaihatsu no Genjo [The State of Japan's Petroleum Development]* (Tokyo: Sekiyu Kogyo Renmei, 1977).

the share of refined products drawn from crude produced by the firm to ten percent. Idemitsu, Taiyo, and others also invested upstream in an attempt to increase scale.[41] These firms remained overwhelmingly focused on refining and marketing domestically, however, and the market remained fragmented.

This lack of scale and risk diversification meant that oil price volatility forced the industry into crisis.[42] This was not because of a loss of physical supplies of crude oil to refiners: estimates in mid-November 1973 suggested a third of contracted import volumes would be lost, however, only twenty percent of contracted crude was lost for a one-month period in November 1973 and February 1974. Otherwise the reduction of import volumes stood at around ten percent of contracted supply. Instead, the crisis emerged because of the short- and long-term effects of price volatility.

In the short-term downstream sector firms faced difficulties because of their inability to pass the increase in the costs of inputs onto consumers. Under the industry law, the ministry had the power to issue guidance over prices to firms. Although industry was allowed to pass on price increases in November 1973 following the announcement of the embargo by OAPEC on October 17, product prices were frozen until March 18, 1974.

In the long term, the changing structure of demand undermined performance. Further, unlike French and many US firms, Japanese firms could not balance losses in the refining sector by gains from the rising value of crude. The thirteen major refining firms recorded combined

41 Staff reporter, "Sekiyu Seisei, Motouri Kakusha, Jimae no Genyu Motome Kaigai Kaihatsu Rasshu [Oil Refiners and Wholesalers Rush into Development Internationally in Search of their Own Crude]," *Asahi Shinbun*, July 15, 1987: 11. Shigeru Kikuchi, Deputy Manager, Overseas Operation Department, Idemitsu Kosan, interview with author, Boston, MA, 2009.

42 More generally, volatility magnified inflation. In the United States, the Nixon administration suspended convertibility of the US dollar to gold on August 16, 1971, and two days later the Bank of Japan revalued the Japanese yen upwards by almost seventeen percent to 308 yen to the dollar, the first adjustment in the yen–dollar rate since the Dodge Line established a rate of 360 yen in 1949. Following this, the Bank of Japan bought US dollars to manage the rise of the yen. Coupled with the loose fiscal policy of Japanese Prime Minister Tanaka, this fed inflationary pressures into the Japanese economy. See Yoshio Suzuki, *Nihon no Kinyu Seisaku [Japanese Fiscal Policy]* (Tokyo: Iwanami Shinsho, 1993), 36–43.

losses of 17.5 billion yen for the first quarter of 1974 in nominal terms.[43] The paid-in capital of firms in the sector as of 1979 stood at 5.2 percent, which compared unfavorably with the average across Japanese manufacturers (21.96 percent) and the US majors (45.57 percent).[44] Although a rising yen meant the cost of crude to refiners fell in relative terms, this was not enough to insulate refiners and wholesalers from increased purchasing costs of crude oil and falling demand for products.

The gap between the oil shocks did not ameliorate the difficult position of downstream firms. In a survey conducted by the regulator, refiners recorded that eight firms had refinery overcapacity of ten percent, with four firms holding up to fifteen percent overcapacity, and five firms with overcapacity of fifteen percent or more.[45] In the financial year 1981, the industry lost 345.6 billion yen, and 123.7 billion in the financial year 1982.[46] The small size and lack of diversification of Japanese firms was identified as the problem. Policymakers noted that poor performance in the downstream sector could not be made up for through crude production; the existence of thirteen wholesalers and thirty-one refining companies meant the market was divided up between small market shares, making it impossible to take advantage of economies of scale.[47]

In exploration and production, rising oil prices initially increased the value of investments. Policymakers increased support for upstream activities by national-flag firms in response to oil price volatility. In 1978 the government vehicle for supporting firms was renamed the Japan National Oil Company (JNOC) and given expanded authority

[43] Nihon Sekiyu, *Nihon Sekiyu Hyakunenshi*, 762.

[44] Sekiyu Bukai, "Sekiyu Sangyo no Keiri Jokyo to Seihin Kakaku Mondai [On Problems Related to the Financial Position of the Petroleum Industry and Petroleum Product Prices]" (Tokyo: Ministry of International Trade and Industry, 1981).

[45] Sogo Enerugi Chosakai, "Kongo no Sekiyu Sangyo no Arikata ni Tsuite (Sanko Shiryo) [Internal Report: On the Future of the Petroleum Industry: Additional Material]" (Tokyo: Ministry of International Trade and Industry, 1981), Table 1–1.

[46] Ministry of International Trade and Industry, "Sekiyu Sangyo ni okeru Kozo Kaizen no Genjo to Kongo no Hoko [The Current State and Future Direction of Structural Reform in the Petroleum Industry]" (Tokyo: Ministry of International Trade and Industry, 1983), 3.

[47] Ministry of International Trade and Industry, "Kongo no Sekiyu Sangyo no Arikata ni Tsuite (Sanko Shiryo) [Internal Report: On the Future of the

to subsidize firms.[48] Equity capital was provided to underwrite exploration, with up to seventy percent of total equity requirements provided for offshore projects, and up to eighty percent for projects on the Japanese continental shelf. If projects reached the development phase, JNOC then guaranteed debt, usually in coordination with private sector banking syndicates that had political risk underwritten by the Export–Import Bank of Japan or the Development Bank of Japan.[49] By 1976, Japanese firms were engaged in forty-four project firms internationally, with the state providing capital to twenty-three of these (fifty-one percent). In 1985 the number of project firms stood at forty-three, with the state investing in thirty-two of these (seventy-four percent). Financing was provided irrespective of the likelihood that the receiving firm would develop into an integrated major, however. As a result Japanese firms' participation in exploration and production remained fragmented. In fact, the structure of support encouraged firms to establish on a project-by-project basis, leading to a proliferation of companies engaged in exploration and production projects internationally.

As with refining and marketing, oil price volatility undermined the performance of firms operating in the upstream. Project financing decisions made by the state assumed that prices would remain at forty to fifty dollars a barrel over the life of the project. This stood in contrast to the international majors at the time, which maintained a twenty-five dollar a barrel assumption.[50] Projects were therefore saddled by debt when long-run oil prices collapsed. An audit of JNOC showed the poor state of upstream investments. The results were presented to the Administrative Oversight Committee of the House of Representatives. They showed that by the end of FY1997 JNOC invested or loaned

Petroleum Industry: Additional Material]" (Tokyo: Ministry of International Trade and Industry, 1981), 3.

48 It also established a national petroleum stockpile in addition to the ninety days of petroleum holdings required to be held by private sector importers.

49 In 1980, the purview of JNOC was extended to allow the purchasing of crude oil and investment in international projects. This was first proposed in the Petroleum Committee interim report of October 23, 1978. See Ministry of International Trade and Industry, "Kongo no Sekiyu Seisaku no Hoko [The Current and Future Direction of Petroleum Policy]" (Tokyo: Ministry of International Trade and Industry, 1978), 14–15.

50 Akira Ishii, Chief Economist, Japan Oil, Gas and Metals National Corporation (JOGMEC), interview with author, Tokyo, Japan, July 5, 2006.

more than 1.7 trillion yen to 266 firms. Of these just forty-four were producing oil, while sixty-eight remained in the exploration phase and ten were in the process of being dissolved. Paper losses reached 342.2 billion yen from the projects invested in by JNOC to the end of 1997, including the 144 firms that had already been dissolved.[51] When asked to explain the losses, Komatsu Kunio, the president of JNOC at the time, noted that oil price volatility had harmed performance: "In the mid-1980s the price of oil collapsed and remained low, and the yen continued to increase in value. These factors had a serious influence on the condition of the companies. I think this was an extremely unfortunate event for our country's upstream petroleum development."[52]

Stable and unified firm strategies

In the last section we saw that, in contrast to French firms, Japanese firms were unprepared to manage the risks and opportunities associated with the change in oil market structure. The argument developed in Chapter 3 suggests that the small size and lack of diversification of Japanese firms operating in refining and marketing, and in exploration and production, should have led them to continue to seek support from policymakers. Discussion on how to adjust the institutions governing oil markets gave firms an opportunity to voice their position to policymakers.[53]

An internal memorandum from 1981 shows that refiners sought support in dealing with overcapacity, and wanted to retain barriers to competition from imports of petroleum products. They also empha-

[51] If prices had not dropped so precipitously, one observer suggested, it is not clear that JNOC reforms would have occurred. Akira Ishii, interview with author. Later calculations determined that the investments held by JNOC were profitable if calculated after the rise in petroleum prices. Nobuyori Kodaira, former administrative Vice Minister, Agency of Natural Resources and Energy (2004–6) and former Director, Petroleum Planning Division (1995–6), Ministry of Economy, Trade and Industry, interview with author, Tokyo, Japan, July 19, 2007; senior official at JOGMEC, interview with author, Tokyo, Japan, July 20, 2007.

[52] Kunio Komatsu, Statement are Administrative Oversight Committee, House of Representatives, December 3, 1997.

[53] Data on the policy preferences of firms are obtained from industry association positions and internal memoranda on firm hearings within committee where available.

sized the importance of maintaining managerial independence from the state.[54]

This opinion was repeated in 1985 hearings. Daikyo Oil noted in its response to a ministerial request for comments that uncontrolled imports would cause significant harm to the national economy and should be avoided, while the refine-at-home policy should be maintained. It also argued that if imports should be allowed, only firms able to ensure the stable supply of products – meaning domestic firms – should be allowed to do so. Daikyo also argued that firms should be allowed to move forward with efforts to create lead firms independently, rather than be forced to merge through state action.[55] Toa Oil similarly argued in the May 30, 1985, hearings that the rationalization of wholesalers should extend to refining in order to improve performance, but that firms should be allowed to determine the pace and direction of this change. Kyodo Oil agreed that firms should be allowed to proceed with the rationalization of overcapacity at their own pace, and proposed that firms receive public support in order to improve their financial positions. Like Daikyo, they argued in prepared remarks that petroleum imports should only be carried out by firms with refining capacity in Japan.[56] The industry association representing the refining industry similarly lobbied the government to maintain the refine-at-home policy; in other words, to limit imports of petroleum products and maintain institutions promoting the interests of domestic refiners, in the name of maintaining stable supply.[57]

Firms recognized the crisis they faced. They agreed to coordinate cutting refining capacity by twenty percent over three years, and drew subsidies to offset the costs of refinery closures. They also agreed to allocate the reduction in market share across the five largest refining firms, and in February 1986, Daikyo, Maruzen, and Cosmo Oil agreed

[54] Sekiyu Shingikai Sekiyu Bukai Shoiinkai, "Naibu Shiryo: Hiaringu ni Okeru Kakushabetsu Iken [Internal Memorandum: Firms' Positions in Hearing]" (Tokyo: Ministry of International Trade and Industry, 1981).

[55] Sekiyu Shingikai Sekiyu Bukai Shoiinkai, "Hiaringu Kaitosho [Hearing Response Form]" (Tokyo: Ministry of International Trade and Industry, June 3, 1985).

[56] Sekiyu Shingikai Sekiyu Bukai Shoiinkai, "Hiaringu Kaitosho [Hearing Response Form]" (Tokyo: Ministry of International Trade and Industry, May 30, 1985).

[57] Sekiyu Renmei, "Sekiyu/Enerugii Seisaku ni Kan Suru Teigen [Proposal Regarding Petroleum and Energy Policy]" (Tokyo: Sekiyu Renmei, 1982), 8.

to merge.[58] Other than Idemitsu Kosan, however, firms continued to argue in favor of retaining barriers to market entry.

This support for ongoing state intervention was repeated in exploration and production. The collapse in oil prices in the 1980s harmed the performance of investments in oil production, and Japanese firms were no exception. As oil prices fell in the 1980s, and the yen rose relative to the US dollar, the industry association noted the distinction between the majors, which were largely to finance operations from retained earnings, and their members, which were reliant on financial support and loans from the state. The association noted that firms were in a parlous state both because projects already in production were producing far fewer revenues, thus making the interest payments from state financing more onerous, and because it harmed the expected commercial value of potential projects.[59]

They continued to demand subsidy support, arguing that projections of falling non-OPEC supply and industry fragmentation meant they relied on the state for capital, in contrast with firms able to cover exploration and production costs using retained earnings.[60] Indeed, the weak position of firms operating internationally meant state subsidies were important to their investments. In 1983, for example, more than forty-six percent of the capital deployed in projects in the production phase came from government, as well as an extraordinary seventy-three percent of total loans.[61]

[58] Staff reporter, "Genyu Seisei Setsubi Sakugen he, Tsusan Hoshin, Ote 5sha ga Doi [Towards a Reduction in Crude Oil Refining Facilities, The Five Largest Firms and MITI Reach Agreement]," *Asahi Shinbun* (June 4, 1985): 9; Staff reporter, "Nakayama Yoshiro Daikyo Sekiyu Shacho Intabyu [Interview with Daikyo CEO Nakayama Yoshiro]," *Asahi Shinbun (evening edition)* (February 22, 1986): 9; Staff reporter, "Sekiyu Seisei Setsubi no Shukusho ni Kikin, Tsusansho, 61-nendo Sosetsu he [MITI Moves to Establish a Fund for Reducing Oil Refining Facilities for FY1986]," *Asahi Shinbun* (August 2, 1985): 9.

[59] Sekiyu Kogyo Renmei, *Wagakuni Sekiyu/Tennen Gasu Kaihatsu No Genjo to Mondaiten [Current State and Problems in Japan's Petroleum and Natural Gas Development]* (Tokyo: Sekiyu Kogyo Renmei, 1986), 6.

[60] Sekiyu Kogyo Renmei, *Wagakuni Sekiyu/Tennen Gasu Kaihatsu No Genjo to Mondaiten [Current State and Problems in Japan's Petroleum and Natural Gas Development]* (Tokyo: Sekiyu Kogyo Renmei, 1987), 18.

[61] Sekiyu Kogyo Renmei, *Wagakuni Sekiyu/Tennen Gasu Kaihatsu No Genjo to Mondaiten [Current State and Problems in Japan's Petroleum and Natural Gas Development]*(Tokyo: Sekiyu Kogyo Renmei, 1983), 19.

Given this, it is little surprise that industry argued both that the government should continue to support their activities, and that changes should be made to government financing in order to make them less onerous. The association proposed, for example, that interest repayments be suspended during the exploration phase. They also proposed changes to the accounting system used to write off debts in order to let them develop multiple projects within a single firm structure, thereby better balancing risks associated with individual projects.[62] The association also sought changes to the structure of financing for oil field development, arguing that firms were facing problems, including country risk, falling oil prices, and a rising yen, that were beyond their capacity to resolve without government support.[63] In requesting these changes, firms linked the serious effects of oil price volatility given their weak position:

> Japanese upstream firms are weak in comparison with the majors in advanced industrialized countries, with fewer retained earnings. They have sunk into a serious situation following the sudden fall in oil prices the year before last, rapid rise in the yen, and falling sales because of oversupply.[64]

The association also repeatedly proposed that the government introduce incentives to domestic refiners to buy their oil.[65]

It is thus unsurprising that – in contrast to the French firms – when policymakers began to address the problem of the poor performance of state investments in oil exploration and development, the industry association representing firm interests argued that firms should be allowed to choose investments independently, but that state support was critical to their ability to compete. Indeed, in public comments submitted to the regulator during debate over how to reorganize the

[62] Sekiyu Kogyo Renmei, *Wagakuni Sekiyu/Tennen Gasu Kaihatsu No Genjo to Mondaiten* (1986), 20.

[63] Ibid., 24–6. This demand was repeated in each annual report of the association through the 1980s and 1990s.

[64] Sekiyu Kogyo Renmei, *Wagakuni Sekiyu/Tennen Gasu Kaihatsu No Genjo to Kadai [Current State and Issues in Japan's Petroleum and Natural Gas Development]* (Tokyo: Sekiyu Kogyo Renmei, 1988), 6.

[65] See, for example, Sekiyu Kogyo Renmei, *Wagakuni Sekiyu/Tennen Gasu Kaihatsu No Genjo to Kadai [Current State and Issues in Japan's Petroleum and Natural Gas Development]* (Tokyo: Sekiyu Kogyo Renmei, 1991), 20.

state subsidies, described below, the industrial association and Toyo Oil both argued that the weak position of firms was due to the way in which state investments had been carried out. The industry association also argued that member firms both needed state subsidies, and expected them to be continue to be provided.[66]

Shifting policymaker strategies

If firms remained poorly diversified and small, and continued to demand support from the state as a result, then why were changes made to Japan's institutions governing oil markets? The answer lies in pressure from manufacturers to lower the costs of inputs, and a shift in the strategy used by policymakers to promote the development of a competitive firm, given the inability of both refiners and firms operating in exploration and production to manage the new market environment.

There is limited evidence to suggest this was driven by changes in the way policymakers understood the nature of risk in the international oil market. Rather, as in the French case, evidence suggests that oil policy was affected by a broader shift in the approach that policymakers took to economic governance. The Maekawa Commission, for example, operated as an advisory body to the prime minister and reported in 1986 that Japan should lower barriers to trade in order to increase imports and domestic consumption. While the Maekawa Commission did not mention energy or petroleum, it noted that imports of manufactured products should be increased in order to improve the international division of labor and make the domestic distribution system more efficient.[67] The Hiraiwa Commission, on the other hand, speci-

[66] Kaihatsu Bukai, "Sekiyu Kodan ga Hoyu Suru Kaihatsu Kanren Shisan no Shori ni Kan Suru Hoshin ni Kakawaru Chukan Torimatome ni Taishite Yoserateta Paburikku Komento [Public Comments Submitted Regarding the Interim Report on the Policy for Disposing of [Petroleum] Assets Held by the Japan National Oil Company] (Tokyo: Ministry of Economy, Trade and Industry," March 10, 2003).

[67] On the use of informal committees see Richard J. Samuels, "Leadership and Political Change in Japan: The Case of the Second Rincho," *Journal of Japanese Studies* 29 (2003), 1–31. For the assessment see Sekiyu Renmei, *Sekiyu Gyokai No Suii [Trends in the Petroleum Industry]*: (Tokyo: Sekiyu Renmei, 1993), 167.

fied that gasoline, kerosene, and light oils should be opened to imports in order to lower costs.[68]

The effects of the changes in the structure of the international oil market on the performance of firms also mattered, as did the political costs of supporting firms also increased as consumers sought to lower prices in order to promote competitiveness. Domestic refining firms were weakened by the price volatility of the first and second oil shocks, as noted above. In October 1980 an internal committee was established to focus on changes in the international market, the effects on domestic industry, and how this should be reflected in national policy. It met eighteen times to May 1, 1981 and had senior representatives from all sections of the ministry involved in oil policy setting, as well as trade officials and representatives from the Cabinet Secretariat. It therefore represents a useful proxy for the status of policymakers' preferences towards oil market governance.[69]

Some members of the committee proposed that the ongoing weakness of Japanese firms demonstrated the failure of the existing strategy, and that it should be jettisoned in favor of the abolition of restrictions on trade in petroleum products, thereby increasing the number of firms qualified to import petroleum products regardless of nationality. Others argued that retaining existing restrictions in the domestic refining market remained the optimal policy. They noted the existing law was the most significant factor ensuring Japanese firms maintained market share within the domestic market. Other arguments marshaled for maintaining this refine-at-home policy focused on the possibility that the open market for petroleum products (as opposed to the more liquid crude market) was inadequate to meet demand, and that keeping barriers to market entry domestically was useful for obtaining information about industry plans.[70]

68 On the history of the deregulation process in the 1980s see Keidanren, *Keidanren 50-Nenshi [Fifty-Year History of Keidanren]* (Tokyo: Keidanren, 1999). See Iwao Nakatani and Hiroko Ota, *Keizai Kaikaku no Bijon: Hiraiwa Ripoto Wo Koete [A Vision of Economic Reform: Beyond the Hiraiwa Report]* (Tokyo: Toyo Keizai Shinposha, 1994), 6.

69 Sekiyu Kihon Mondai Kenkyukai, "Sekiyu Kihon Mondai Kenkyukai Chukan Hokokusho [Interim Report of the Research Committee on Fundamental Problems in Petroleum]" (Tokyo: Ministry of International Trade and Industry, 1981).

70 Sekiyu Bukai, "Naibu Shiryo: Kongo No Sekiyu Sangyo No Arikata ni Tsuite [Internal Report: On the Future of the Petroleum Industry]" (Tokyo: Ministry of International Trade and Industry, 1981).

Debate was also prompted by consumer demands. The position of major consumer industries within Japan shifted in response to change in the oil market. In the 1970s, fears about disruptions to physical supply meant major consumers supported increased strategic intervention on energy security grounds. The burden this intervention placed on individual consumers of gasoline, and by extension the auto industry, was recognized for the first time in the main policymaking body, however. In 1984, consumer preferences supported lowering prices.[71] The report from the peak energy decision-making body noted that the "rise in energy costs has had a significant effect on basic industries and broadly across Japanese economy and society. The reduction of energy costs is therefore an important issue for the national economy."[72]

Further evidence in support of the contention that consumer preferences had shifted can be found in the position adopted by peak industry body Keidanren. There was little initial opposition to the increase in intervention by the state from major energy consumers. Keidanren agreed that subsidies should be increased in order to strengthen the performance of the sector.[73] As oil prices fell, however, it argued that the best solution to the problem was to let the market set prices during normal market operation but retain emergency powers.[74] By doing so, it argued, consumers could take advantage of lower energy costs, while also providing insurance against supply problems in the case of

[71] Ministry of International Trade and Industry, "Sekiyu Ryutsu Bijon Kenkyukai – Chukan Hokoku: Shiryo 1: Shohisha Ni-zu ni Kansuru Giron Pointo [Interim Report of the Research Committee for a Vision for Petroleum Distribution: Materials 1: Summary of Debate on Consumer Needs]" (Tokyo: Ministry of International Trade and Industry, 1984).

[72] Cited in "Sekiyu Sangyo no Kozo Kaizen no Hoko to Sekiyu Sangyo Seisaku no Arikata ni Tsuite [Structural Reform of the Petroleum Industry and the State of Petroleum Industry Policy]" (Tokyo: Ministry of International Trade and Industry, 1984), 15. Rising consumer demand for cheaper inputs is also noted in Staff reporter, "Sekiyu Seihin Yunyu he: Shoin, Tsusansho, Shingikai ni 3-gatsumatsu Secchi; Yushutsu Danryokuka mo Kento [Towards Oil Product Imports: Small Committee within MITI Established in March; Also to Explore Loosening Exports]," *Asahi Shinbun*, March 10, 1985: 9.

[73] Keidanren, "Sekiyu Wo Chushin to Suru Enerugii ni Kansuru Yobo [Request in Relation to Petroleum and Other Energy Resources]" (Tokyo: Keidanren, 1973).

[74] The positions of Keidanren and the industry association matched on the question of tax increases, on the other hand. Keidanren argued that an increase in the petroleum tax was unjustified, and repeated its position that energy security of supply policies should be funded out of general revenues, given

market disruptions. In a May 1988 statement, Keidanren included petroleum, along with other energy sources, as an area to be included for ongoing liberalization and deregulation.[75]

Evidence shows that policymakers remained committed to supporting domestic industry.[76] There was, therefore, no consensus on jettisoning support for domestic firms.[77] The solution proposed was instead to encourage consolidation in the industry:

> The central axis of the petroleum supply system in Japan's domestic petroleum industry is the wholesalers, who obtain supply of crude oil, produce [refined products] and then distribute. Increasing the concentration of wholesalers is extremely important in order to establish a supply platform that is stable and efficient ... Under the current extremely difficult conditions, it is more of an urgent task than ever to increase industry concentration among wholesalers in order to respond quickly and appropriately to the needs of the national economy through the stable supply of petroleum.[78]

The change to oil market governance finally negotiated between business and the regulator committee responsible for designing oil policy enabled greater imports while maintaining support for the domestic industry. This, it was deemed, would allow entry and exit into the market to remain slanted in favor of Japanese firms. Allowing prices to operate freely was expected to force mergers within industry, and therefore move industry closer to producing a national-flag firm capable of competing in the international petroleum market. Some twenty years later, the head of the petroleum section in the competent ministry

their status as national policy. Sekiyu Renmei, *Sekiyu Gyokai No Suii [Trends in the Petroleum Industry]* (Tokyo: Sekiyu Renmei, 1986), 144–5.

75 Keidanren, "Kisei Kanwa Ni Kansuru Yobo [Request Regarding Deregulation]," in *Keidanren 50-Nenshi [Fifty-Year History of Keidanren]*, 557.

76 Ministry of International Trade and Industry, "Sekiyu Sangyo No Kozo Kaizen No Hoko to Sekiyu Sangyo no Arikata ni Tsuite [On the Direction of Structural Reform in the Petroleum Industry and the State of the Petroleum Industry]" (Tokyo: Ministry of International Trade and Industry, 1985), 11.

77 Ministry of International Trade and Industry, "Sekiyu Sangyo No Kozo Kaizen No Hoko to Sekiyu Sangyo no Arikata ni Tsuite," 21.

78 Ministry of International Trade and Industry, "Motouri Kigyo no Shuyakuka ni Tsuite Chukan Hokoku [Interim Report on the Concentration of Wholesale Firms]" (Tokyo: Ministry of International Trade and Industry, 1984), 3.

at the time, Masahisa Naito, noted the intentions within the ministry at the time: "deregulation is nothing more than an instrument. In other words, through deregulation [I wanted to] increase industry concentration. Why increase concentration? In order to make an oil firm that is able to compete internationally."[79]

The new approach to managing markets was accepted by legislators in the ruling Liberal Democratic Party.[80] Special legislation passed through parliament that operated alongside the existing law, and allowed imports of gasoline, kerosene, and light oils, but limited trade to firms with refining and storage capacity within Japan and able to meet Japanese quality standards.[81] In practice this meant that only Japanese refiners and wholesalers were able to meet the requirements.[82] National control therefore remained the leitmotif of intervention in the petroleum sector. The law included a ten-year sunset clause, leaving open the possibility that trade could be restricted once again if the balance of supply and demand in crude oil became less favorable in the 1990s.[83] The law ensured that both foreign and Japanese competition were banned from importing petroleum products.

[79] Interview with Masahisa Naito, *Enerugi Foramu* (September 2005): 62; This is corroborated by Tsutomu Toichi, Managing Director, Institute Energy Economics Japan and Petroleum Council member, interview with author, Tokyo, July 4, 2006; and Nobuyori Kodaira, interview with author. See also Staff reporter, "Sekiyu Seihin Yunyu he: Shoin, Tsusansho, Shingikai ni 3-gatsumatsu Secchi, 9.

[80] Staff reporter, "Sekiyu Seihin Yunyu Hoan, Torokusei de Jimin Ryosho [LDP Agrees to Oil Produce Import Bill with Registration System]," *Asahi Shinbun*, October 24, 1985: 9.

[81] Under the existing law imports were not banned, but rather importers were required by law to register import plans with MITI. Under this system MITI had only recognized the importation of naphtha, LPG, and heavy oil in order to enable firms to balance supply and demand, but imports of gasoline, kerosene, and light oils were banned.

[82] Critics of the bill argued that it increased, rather than decreased, the extent of regulatory control over trade in petroleum products, establishing in law a practice that had until then been carried out through informal guidance. Yasuo Fukuoka, Trade and Commerce Committee, 103nd Session of the Diet, Japanese House of Representatives, November 20, 1985; Hiroshi Takahashi, "Gasorin Yunyu Kaikin: Matta Nashi no Ikinokori Kyoso Jidai he [Allowing Gasoline Imports: No Escape from the Era of Competition for Survival]," *Shukan Toyo Keizai*, February 8, 1986: 90–3; Nobuyori Kodaira, interview with author.

[83] The bill originally proposed to place a five-year limit on the right to import, based on IEA predictions of a tight market in the 1990s, but this was extended

A 1996 review confirmed the benefits of allowing imports, and concluded that intervention was unnecessary. The law expired, therefore, marking the abandonment of attempts by the state to increase strategic control of petroleum through the protection of firms engaged in domestic refining, and leading to widespread reorganization of the industry, including mergers between the major Japanese refining interests and foreign majors.[84] Gradual market opening was implemented over a five-year period. The licensing of refining capacity was loosened immediately. The regulation of firms' production plans was abolished by the end of 1988, and the regulation of firm refining schedules was abolished within five years. A member of the committee established to design these changes, Yoshihiro Morozumi, noted that the changes were "epochal for petroleum policy."[85]

It is worth noting that the reduction of regulatory and trade restrictions from 1986 to 1996 had the desired effect. After 1996 the price spread between the international and domestic markets for petroleum products lessened. Downstream firms shifted to a new pricing regime for gasoline, kerosene, and light oil that adjusted monthly to reflect international market prices.[86] The controlled introduction of imports of gasoline and other fuels also led to firm mergers. In expectation of the change, Showa Oil and Shell merged with one another in 1995. Then, following the entry into force of the law in 1996, Cosmo, Daikyo, and Maruzen joined together under the Cosmo banner, followed by Nihon Kogyo and Kyodo Oil in 1992, which formed Japan Energy. A second wave of mergers began after the abolition of the special law. The largest downstream refiner, Nihon Sekiyu, cut its longstanding ties with Caltex and merged with Mitsubishi Oil in April

to ten years during Diet interpellations, under the rationale that a review of market conditions after five years was premature.

84 The change in government during 1993–4 also weakened the veto power of those arguing to maintain a ban on imported products. Kosuke Oyama, "The Policymaking Process Behind Petroleum Industry Regulatory Reform," in *Is Japan Really Changing Its Ways? Regulatory Reform and the Japanese Economy*, eds. Lonny E. Carlile and Mark Tilton (Washington, DC: Brookings Institution Press, 1998), 142–62.

85 Sekiyu Sangyo Kihon Mondai Kentokai, "Sekiyu Sangyo Kihon Mondai Kentokai Hokokusho [Meeting Minutes of Report of the Deliberative Committee into the Petroleum Industry]" (Tokyo: Ministry of International Trade and Industry, 1988).

86 Sekiyu Renmei, *Sekiyu Gyokai No Suii [Trends in the Petroleum Industry]* (Tokyo: Sekiyu Renmei, 1996), 3.

1999, then entered into partnership in refining and distribution with Cosmo Oil in October of the same year. Japan Energy and Showa/Shell also joined refining and distribution networks, although they did not formally merge. Then, in 2000, Toa and General Oil merged to form Tonen-General, and finally in 2002 Idemitsu and Shin-Nihon Sekiyu joined refining operations. By the end of the process firm mergers and partnerships reduced the sixteen firms operating in the downstream sector to four groups.

Oil price volatility also harmed the position of firms operating upstream, and led to a review of the subsidy program for exploration and production internationally. Given ongoing support among both firms and policymakers for intervention, however, this review led to a reorganization rather than an abolition of the program of state subsidies. Low oil prices in the 1980s and 1990s hurt returns across the oil industry. Japanese firms operating upstream suffered because of their inability to manage risks on a portfolio basis. Instead, fragmentation in the industry meant it was in a poor state when judged on a project basis. The competent minister found that just eleven of 123 firms invested in by the government were nominally in the black, with the number likely to rise to nineteen firms by 2020. Of the 1.392 trillion yen in subsidies distributed to the end of 1997, total unrecoverable investments stood at 438 billion yen after the sale of assets and repayment of investments from profitable firms.[87]

A second review focused on reforming the subsidy program in order to better promote the development of an integrated energy firm. It proposed an increase in state intervention for five years, allowing JNOC to buy stakes in existing fields in order to develop the portfolio of the proposed major. Firms would then continue to receive support for five years, before government withdrew from subsidy provision.[88] The industry association, on the other hand, rejected the view that state support

[87] Statement by Kaoru Yosano, Minister for International Trade and Industry, on October 13, 1998, to the Economy and Industry Committee, House of Representatives, 142nd Diet. The revenue flows were calculated under two scenarios. The first used a \$16.1/barrel Dubai assumption (the lowest value from the average spot price of the last 3–5 years), with an exchange rate of 110 yen/dollar (five-year average). The second used a \$20.7/barrel (highest six month average from the last six years) and 145 yen/dollar (monthly average for August 1998) assumption.

[88] Interviews confirm the committee's intention to withdraw state support after a ten-year window, citing the example of the German firm Deminex, which was

for upstream activities should be temporally bounded, and that policy should "pick winners" by channeling subsidies only to firms judged capable of growing to a scale enabling them to compete internationally.

Changes to the structure of subsidies were institutionalized through the abolition of JNOC and the transfer of many of assets to a new public body, the Japan Oil, Gas and Metals Mining Agency of Japan (JOGMEC). The new body differed because its portfolio of assets was reduced. Companies in which JNOC held shares were broken into three groups. The first was merged and the new company listed on the stock exchange.[89] The second was sold off through tender.[90] The third set of assets – those of firms not yet in the development phase – remained in JOGMEC hands. This sell-off of assets, recommended in 2000, had its first success with the emergence of a new and more powerful upstream player, created through the merger of the former JNOC-financed INPEX and Imperial Oil. The newly listed INPEX, which operated internationally, and Teikoku Oil, which owned mostly domestic assets, announced their merger in December 2005. Although small on international terms, the firms, which had been the number one and number three firms in terms of upstream assets, were predicted to reach a combined production of 560,000 barrels of oil equivalent by the financial year 2009.[91] In an interview with business daily the *Nihon Keizai Shimbun*, the chairman of the newly merged firm, Matsuo Kunihiko, noted that the firm aimed to grow by five percent per year, and to reach a production level of one million barrels by

dissolved by the government in 1997. Mitsuo Arai, member, Committee on Basic Problems in Petroleum (2000) and member, Petroleum Council (2001–2007), interview with author, Tokyo, Japan, July 12, 2007; Takeo Kikkawa, member, Committee on Basic Problems in Petroleum (2000) and member, Petroleum Council (2001– present), interview with author, Tokyo, Japan, July 18, 2007.

89 The assets of INPEX and its subsidiaries, along with JODCO, were used.

90 Lead companies in these firms were given first bidding rights on the projects, and by February 2005 twenty-four of twenty-seven firms had been sold off for a total profit of 13.49 billion yen. For a complete list see Sekiyu Kogyo Renmei, *Wagakuni Sekiyu/Tennen Gasu Kaihatsu No Genjo to Kadai [Current State and Issues in Japan's Petroleum and Natural Gas Development]* (Tokyo: Sekiyu Kogyo Renmei, 2006), 59–60.

91 Staff reporter, "Jun-meja Mezashite Keiei Togo – Sogo Hokan to Kibo Kakudai de Kaigai Keneki no Kakutoku he [Merger Aimed at Creating a Mini-Major – Looking to Secure International Development Rights through Mutual Support and Scale Growth]," *Energy* (December 2005): 18–19.

2015, making it one-seventh the size of market leader Exxon-Mobil in terms of its reserves base.[92]

The second change saw lending conditions to the private sector restricted. The provision of loans was halted, and equity capital supplied for overseas exploration projects reduced to fifty from seventy percent of total capital costs. Loan guarantees were also cut to fifty percent,[93] making government financing far less attractive to firms. By 2006 JOGMEC had offered no new loans or guarantees to upstream firms' exploration projects. Although high oil prices contributed to this, the upstream industry suggested this was also driven by the restrictive lending conditions, and continued to lobby for JOGMEC to extend funds not only to firms with the ability to grow into a Japanese major, but also to other firms.[94]

Summary

Prior to the 1980s, Japanese firms were small and poorly diversified. The industry was split between refiners and marketers focused wholly on the domestic market, and firms engaged in exploration and production internationally, and there were numerous firms operating in each of these segments. This stands in contrast with the French case, although firms in both countries began under similar initial conditions, with a lack of domestic resource endowments. Japanese firms also successfully gained access to production opportunities internally, especially from the 1950s, making it implausible that the difference in firm characteristics can be attributed to the French colonial possessions with oil reserves.

Instead, in this chapter I presented evidence that suggests differences in the characteristics of firms between the two cases were shaped by the industrial compact negotiated between business and government in each. In Japan, domestic industry was protected through trade barriers, but laws did not encourage mergers between firms. Instead, they encouraged firms to compete over refining capacity, and firms rarely

[92] Staff reporter, "Intabyu: Kokusai Sekiyu Kaihatsu Teikoku HD Kaicho Matsuo Kunihiko [Interview: Matsuo Kunihiko, chairman of INPEX-Teikoku Oil]" *Nihon Keizai Shinbun*, December 28, 2005: 7.

[93] JOGMEC, *Heisei 17-Nendo Keikaku [Corporate Plan for 2005]* (Tokyo: JOGMEC, 2004), 11.

[94] Toshiaki Tanaka, former employee, Mitsubishi Oil, and member, JOGMEC Technical Committee, interview with author, June 11, 2006.

sought to increase scale by merging with others. In exploration and production, subsidies encouraged firms to enter the oil business, but also to remain small, rather than increase in scale. Together, the institutions governing the Japanese oil market supported firms, but did not encourage them to merge into a vertically integrated major able to compete on international markets. Firms remained internationally uncompetitive as a result.

The small scale and lack of diversification meant firms operating in the domestic market continued to seek support from policymakers in the 1980s and 1990s. Almost all refiners lobbied to retain laws limiting imports of oil products and regulations that limited the right to imports to firms with domestic refining capacity and storage facilities. These inevitably tilted the competitive landscape in favor of domestic industry. In the exploration and production phase, the small scale of firms and lack of integration meant they were unable to balance price, geological, geographic, and other risks. They therefore also continued to support intervention by the state in the form of subsidies and low-interest loans, in order to reduce the risks to which they were exposed.

Why, then, did partial liberalization occur from the 1980s? This chapter has suggested that the primary cause was the effects of the shift in oil market structure on policymakers' calculations of the costs and benefits associated with maintaining the status quo, and pressure from manufacturers. In the 1970s, the profitability of Japanese firms operating in the domestic sector was undermined by the shift to a high-price environment. Policymakers with responsibility for petroleum policy initially advocated heavier intervention in the petroleum market. In the long run, however, longstanding protection of domestic firms on strategic grounds was abolished. Two changes, in particular, led to this change. First, policymakers decided that trade liberalization could force mergers in the industry; liberalization therefore offered an alternative instrument through which to promote firm strength. Second, consumers seeking to lower the costs of inputs lobbied for liberalization of oil product markets in order to promote their competitiveness, similarly to the French case.

Adjustments made to industrial compact governing in the upstream similarly reflected the shifting strategy of policymakers. Price volatility undermined the financial position of the government vehicle used to provide capital and risk-hedging services to private sector firms. The existence of losses, by themselves, did not force reform of upstream policy. Unlike in the domestic market for refining, the costs of failure in exploration and production did not weigh on oil-consuming industries, and there was little pressure from these groups to reform the structure of intervention. But the existence of swathes of red ink on the books of JNOC interacted with a more general privatization movement within government; a shift in the strategy used to promote economic competitiveness across the industry as a whole.

Partial liberalization thus did not amount to a rejection of state support. Instead, it represented an adjustment to the existing instruments used to pursue this goal after they were undermined by the effects of shifts in long-run oil prices. It is unsurprising, therefore, that ongoing support for national control by policymakers, as well as the weakness of Japanese firms in the international oil market, meant the government continued to offer improved terms to national-flag firms operating in the upstream.

Alternative explanation – ideational change

In Chapter 3 I proposed that ideational change represents an alternative explanation for outcomes. As in the French case, the broad shift towards liberalization also affected outcomes in the Japanese oil market. If this best explains outcomes, I proposed, then we should expect policymakers to make decisions relatively independently of other socioeconomic actors, including firms. Further, to the extent that policymakers are the most important actors in the design and implementation of policies, we should expect them to make changes to the institutions governing oil markets because they have learned that changes to the international oil market have reduced the risks associated with relying on the international market for the supply of oil. Finally, cross-national variation in outcomes should reflect variation between policymakers over the assessment of risk in the international oil market.

The first of these hypotheses – that policymakers' institutional choices determined the form of governance in oil markets – is partially confirmed by the data presented above. The preferred policies of firms operating in refining and marketing, and in exploration and production internationally, matched those predicted in Chapter 3. Japanese firms remained comparatively small and poorly diversified throughout the period covered in this study. Firms operating in both segments continued to opt for non-market strategies, lobbying the government to protect their share of the domestic market and provide capital to help them expand operations internationally.

Given the stability in firm preferences, the changes that were made to the industrial compact used to govern oil markets in Japan were thus led by policymakers rather than firms. The most important question in assessing the explanatory power of the learning hypothesis therefore lies in determining what led to the change in policymaker strategies towards intervention in the oil market. Here there is some evidence to support the contention that learning about the changed nature of risk in the international oil market drove the changes in policymaker strategy. In an assessment about the price volatility produced in 1994, for example, the main specialist committee charged with reassessing state policy towards the petroleum industry noted that:

> The development of the international oil market has positively affected Japan's ability to ensure stability of oil supply through the dispersal through the market of the influence of decreases in supply, and other effects ... An emphasis on reliance on the functioning of the international oil market should be at the basis of policy going forward ... Regarding the refining industry, supply and demand adjustment mechanisms in the petroleum industry law should be abolished, as should regulations allowing for price intervention in peacetime.[95]

On balance, however, the weight of evidence suggests that changes made to the institutions governing oil markets reflected a shift in the strategies adopted by policymakers, and to a lesser degree firms, as they sought to reduce the negative effect these institutions had on firm performance.

Two other calculations were apparent in the change in policymaker sentiment towards oil market governance in Japan. First, the relaxation of barriers to trade in oil products also reflected the trade-offs policymakers faced once consumers began to call for lower energy prices in order to improve competitiveness. Second, as we saw in the last chapter, the changes implemented to policies in the upstream partially reflected changing sentiment towards the perceived benefits of

[95] Sekiyu Bukai, "Sekiyu Seisaku Kihon Mondai Shoiinkai: Chukan Hokoku [Interim Report: Committee on Basic Problems in Oil Policy]" (Tokyo: Ministry of International Trade and Industry, June 20, 1994).

state ownership of productive assets; the abolition of JNOC was one component of an attempt by policymakers to reduce state intervention in the economy more broadly, and therefore represented a shift in the national strategy to promote growth and competitiveness across the economy. On the other hand, the insertion of JNOC into the privatization debate that occurred in Japan came as a direct result of the poor performance of the assets invested in by the state. Further, there is little evidence that policymakers rejected the idea of promoting a national champion by providing capital to private sector firms. In short, the weight of evidence supports the contention that it was the change in the oil market, and the effect that it had on firms, that was the most important cause of institutional adjustment in the case of Japan.

In Chapter 6, I examine the politics of oil in the United States. The United States complements the Japanese and French cases in two ways. First, in terms of firms, the structure of the US industry shows within-case variation, with both major international oil firms in the form of Exxon and Chevron, which are vertically integrated and have substantial international operations, operating alongside small- and medium-sized firms that have little international presence and focus on particular segments of petroleum production, as well as refiners focused on producing products for the domestic market. It thus enables us to examine the effect of firm characteristics on their strategy towards government while holding national political institutions constant. The United States also offers a useful extension of the argument developed above: unlike Japan and France, the United States has substantial domestic resources within national boundaries. Extending the analysis to the United States therefore enables us to examine how industrial compacts negotiated between business and government interacted with domestic resource endowments to shape the characteristics of firms, and therefore the strategies they took towards government.

6 *The United States and energy independence*

As the world's No. 1 consumer, we are forced to buy one-third of our oil from outside the country. We are uncomfortable with a supply and a price history that is out of our hands.

Statement by Theodore A. Burtis, chairman of Sun Company, Inc., to the Senate Energy and Natural Resources Committee, March 20, 1986

In the last two chapters I examined how differences in the characteristics of firms in France and Japan affected the demands they placed on government, and how this affected the laws and regulations used to govern oil and oil products in these countries. I argued that differences in the characteristics of firms in these countries systematically affected the types of demands they made to policymakers, and that these demands influenced the way governments chose to oversee oil markets in these two countries. I also argued that the characteristics of firms in these countries were shaped by a series of what I called "industrial compacts": agreements negotiated between firms and governments in earlier periods, which regulated market share, and shaped the organizational structure and incentives of firms operating in the sector.

In this chapter I extend the analysis of oil market governance to the United States. At first glance the case of the United States appears at odds with the French and Japanese cases. An important factor common to both the French and Japanese cases is the lack of domestic oil resources, making domestic resource endowments an implausible explanation for differences in the characteristics of firms that emerged in these two countries. In contrast, the volume of oil produced in the United States far exceeded that of other countries prior to the emergence of the Middle East as an important center of production. This makes initial conditions in the United States quite different from both France and Japan.

The case of the United States is nevertheless useful for two reasons. First, while examining the causes of changes in oil market governance in France and Japan enabled us to discount the importance of resource endowments in shaping firm characteristics, there are differences in the political institutions of these two countries that could plausibly affect the characteristics and behavior of firms. Certainly both Japan and France are often characterized as having relatively close relations between business and government, which allows us to link strategies to outcomes. Yet beyond this there are differences in political institutions in the two countries that could conceivably affect the strategies adopted by firms, and this is not something we can control for through the comparative research design. In the United States, on the other hand, there are important differences in the characteristics of firms operating in the oil sector, with a division between oil majors that are large and are diversified both internationally and across the supply chain, and smaller producers that operate largely within the United States.[1] This difference allows us to investigate how the strategies adopted by firms are shaped by their characteristics unencumbered by differences in national political institutions.

The case of the United States is useful for a second reason. In the previous two chapters I showed how industrial compacts negotiated between business and governments shaped the characteristics of firms operating in France and Japan, arguing that the differences between firms then went on to influence the strategies these firms adopted in response to the transformation of the international oil market. Extending the analysis to oil market governance in the United States allows us to consider how the existence of domestic endowments of oil interacts with these compacts in shaping the characteristics and behavior of firms. In general terms, scholarship shows that countries that are relatively well endowed in natural resources, including oil, tend to

[1] The International Energy Agency (IEA) notes that "the primary characteristic of the US oil industry is that it consists of a large number of very heterogeneous firms. Nearly 8,000 companies produce oil and/or gas; some 7,400 of them are very small, privately held operations, many producing only a few barrels a day. The majority of the publicly traded companies produce less than 10,000 b/d. About a dozen companies are generally classified as oil majors." International Energy Agency, *Energy Policies of the IEA Countries: The United States 1998 Review* (Paris: IEA, 1998), 50. A comprehensive survey of the US oil industry carried out in 1950 shows similar industry characteristics. John G. McLean and Robert Haigh, *The Growth of Integrated Companies* (Norwood, MA: Plimpton Press, 1954), especially 36 and 48.

be net exporters, and this was certainly the case for the United States prior World War II.[2]

The general statement that resource abundance leads countries to become net exporters does not tell us, however, about the types of firms that emerged to produce and trade this oil.[3] In fact, the existence of domestic oil resources mattered to the emergence of large diversified firms in the United States, as I describe below. Yet, domestic oil reserves also enabled numerous small domestic producers to exist. Further, while managerial skill undoubtedly explains the continued success of many of these firms, they have also benefitted from the subsidies provided by government. At times they have also gained more substantial benefits. The rise of non-US-produced oil in the 1950s, for example, harmed the interests of independents operating predominantly within US borders. For these firms, the diversification of supply increased import competition, other than for a small number of independents that managed to take advantage of the diversification of supply globally to increase their investments internationally.[4] Their response was to lobby for protection from the government, and congressional testimony suggests that a large number of these firms would have failed without the protection provided by government. One effect was to protect the economic interests of smaller producers, which ensured they remained an important interest group in future debates over how to govern the oil market.[5]

The differences in the characteristics of firms meant these firms also responded differently to the transformation of the international oil

[2] Edward E. Leamer, *Sources of International Comparative Advantage: Theory and Evidence* (Cambridge, MA: MIT Press, 1984).

[3] Research has now established that firm heterogeneity is the norm across industries, even in industries in which countries are understood to have a comparative advantage. For a summary see Andrew B. Bernard, J. Bradford Jensen , Stephen J. Redding, and Peter K. Schott, "Firms in International Trade," *NBER Working Paper No. 13054* (April 2007).

[4] Landeau, for example, identifies twenty-one independent firms that shifted operations internationally following World War II. A survey of annual reports from the firms finds the competitive threat of imports, and the greater profitability of non-US production, were important factors in driving international investment decisions. See Jean-Francois G. Landeau, *Strategies of US Independent Oil Companies Abroad* (Ann Arbor, MI: University of Michigan Press, 1979).

[5] A general statement of this type of feedback effect can be found in Paul Pierson, "Review: When Effect Becomes Cause: Policy Feedback and Political Change," *World Politics* 45 (1993): 595–628.

market, consistent with the French and Japanese cases. For the majors, Middle Eastern oil was a bonanza so long as they retained control over its production, and congressional testimony and data drawn from annual reports suggest they were less supportive of the imposition of trade barriers as a result. They responded to the loss of reserves, as in France, by seeking to diversify operations. This included promoting production within the United States, even as they opposed the imposition of barriers to trade. Smaller and less diversified firms, on the other hand, remained focused on the domestic market, and continued to seek the support of the US government in the form of subsidies. Some also sought the reintroduction of import barriers in response to the low-price environment of the 1980s and the 1990s.

The argument developed in Chapter 3 suggests that while the characteristics and strategies adopted by firms operating in the oil sector are important in explaining patterns of oil market governance, it is also important to take into account the incentives for policymakers to provide the support that firms demand from them. If we assume that an important motivation of policymakers is reelection, then in the case of the United States, as in France and Japan, the political implications of oil prices provide an incentive for both the president and Congress to pay attention to the structure of oil market governance.

In contrast to France and Japan, promoting domestic production makes sense for many policymakers because growth in domestic production is associated with state- and district-level benefits.[6] The United States ranks first globally in terms of total wells drilled, and is the third biggest producer globally. In 2004 the Independent Petroleum Association of America (IPAA) claimed there were 1.7 million people employed in the petroleum industry.[7] Given this, we should expect policymakers from oil-producing districts and states to have strong and stable preferences in acceding to the demands of domestic firms in order to represent the interests of their communities and increase the likelihood of reelection. Important congressional committees focused on oil policy have significant representation from oil-producing states. In the Energy and Commerce Committee of the House of Representatives, which is commonly assigned parts of energy legislation proposed in

[6] Bruce Ian Oppenheimer, *Oil and the Congressional Process* (Lexington, MA: Lexington Books, 1974).

[7] Independent Petroleum Association of America, *IPAA Oil and Gas Producing Industry in Your State* (Washington, DC: IPAA, 2005–6).

Congress, members from oil-producing states constituted a mean of seventy-nine percent of the committee membership between the 96th and 109th Congresses (1981–2006). Congressmen from oil-producing states made up seventy-four percent of committee membership over the same period in the House Ways and Means Committee, which considers fiscal instruments to support the industry. The status of oil-producing members in the Senate similarly has broad membership from oil-producing regions. Members from oil-producing states made up eight out of ten of the representatives on the Senate Committee on Energy and Natural Resources between the 103rd and 109th sessions of Congress (1993–2006), as well as on the Senate Finance Committee.[8]

This does not mean we should expect members to be wholly captured by oil interests, nor that the interests of the oil industry are the only important consideration for congressional committees. Congressional members in oil-producing regions are likely to balance the local industry against general voter opinion. The broad range of groups with interests in energy also means bills are commonly split between multiple committees, diluting the influence of committee members from oil-producing districts. It also means that the trade-offs between economic interests and support for the industry are negotiated between members from producer districts and those for whom the oil industry is not an important part of district or state interests. The domestic production of oil and natural gas within national borders means, for example, that there are important interest groups that care about the environmental effects of oil exploitation, and this matters for debate over federal lands and the Outer Continental Shelf (OCS). Indeed, the congressional record suggests that the most important battles over increasing access to federal lands and the OCS revolve around oil firms, which I show below are united in support of increasing access, and environmental interests, who tend to oppose greater access to these areas.[9] Thus we should not expect the collective decision of

[8] Author calculations from Garrison Nelson, *Committees in the US Congress, 1947–1992*; Charles Stewart III and Jonathan Woon, *Congressional Committee Assignments, 103rd to 112th Congresses, 1993–2011*. Statistical reviews probing the relationship between committee assignments and constituency interests find mixed results, but are hampered by data challenges. Qualitative data suggests an association. For a review see Gary W. Cox and Mathew D. McCubbins, *Legislative Leviathan: Party Government in the House*, 2nd edn (Cambridge University Press, 2007), 21–32.

[9] I thank Charles Ebinger for emphasizing this to me.

Congress to lead it to wholly accede to the demands of firms. The evidence below nevertheless suggests they win enough to ensure limited support continues to flow to domestic industry.[10]

Oil market governance to 1980: between conservation and production

The oil industry in the United States is more diversified than in Japan and France. Industry data show that the modal firm is small and domestically focused. A 1996 survey by the IPAA records that fourteen percent of member firms operated internationally, and that slightly over twenty percent of firms raised capital from equity markets, while thirty-four percent raised capital from internal sources; banks provided the next largest tranche of financing, at twenty-six percent of member firms.[11] This contrasts with the US majors, which made up eight of the twenty largest firms in the US Fortune 500 in 1980, with average revenues of more than thirty billion dollars.[12] The largest of these firms is the former Standard Oil (New Jersey) – today called Exxon – with revenues of seventy-nine billion dollars in 1980, as well as 7.3 billion dollars of capital assets in the United States and 5.1 billion dollars held in non-US markets. These assets yielded 560,000 barrels of oil production daily internationally, against 787,000 within the United States. The same year its return on capital stood at 36.7 percent

[10] Policymakers have a second reason for seeking to shape outcomes in oil markets: public support for energy independence. Uniform time series data on the public opinion of energy independence are unavailable, making it difficult to precisely determine how the public popularity of energy independence has changed over time. Yet a wealth of data is available that asks, in different ways, about the utility of increasing domestic production in order to increase energy independence. Together it suggests the popularity of energy independence among the public. Questions about energy independence take various forms between surveys, making it difficult to isolate public opinion from the effects of framing. A review of all survey research recorded with the iPoll Databank of the Roper Center for Public Opinion between 1978 and 2005 that include the term "energy independence" suggests that public support for energy independence has been stable across time, including after the price of crude oil began to fall in the 1980s. It also suggests this support is tempered by environmental concerns.

[11] Independent Petroleum Association of America, *1996 Profile of Independent Producers* (Washington, DC: IPAA, 1996).

[12] Exxon, Mobil, Texaco, Chevron, Gulf Oil, Amoco, Conoco, Sonoco. Data drawn from Fortune 500 ranking.

internationally, as opposed to 29.2 percent within the United States. Exxon also maintained significant refining operations outside the United States, with earnings of 1.7 billion dollars internationally, compared to 202 million dollars within the United States.[13] Conoco, one of the smaller majors, was also diversified internationally, producing 133,785 barrels within the United States, compared to 228,793 barrels per day internationally from fields in the North Sea, Africa, the Asia-Pacific region, and Latin America.[14]

What is the source of this firm heterogeneity? In order to answer this question I begin, as in the previous two chapters, by summarizing oil market governance in the United States. My goal is not to record a complete history of business and government relations in the US oil industry,[15] rather it is to consider how the industrial compacts negotiated between business and government, and the presence of domestic resources, interacted in shaping the characteristics of firms operating within the United States.

An important difference between the US and French and Japanese cases, as noted above, is the existence of significant volumes of domestically located oil. Oil was produced in greater volumes in the United States than any other country prior to the rise of production in the Middle East. An important problem facing firms seeking to profit from these resources replicated that faced by firms operating outside the United States: how to limit supply in order to avoid the problem of overproduction.[16] Initially there were few rules regulating production in new and existing fields, and small operators exploited finds in Pennsylvania, Ohio, and West Virginia with little concern for waste, with populations moving to a new location as fields were exhausted.

[13] Exxon Corporation, *Exxon Incorporated Annual Report* (New York: Exxon Corporation, 1980).

[14] Conoco Incorporated, *Conoco Incorporated 1980 Annual Report* (Stamford, CT: Conoco Incorporated, 1980).

[15] For a useful general history of the early development of the industry see Raymond Foss Bacon and William Allen Hamor, *The American Petroleum Industry* (New York: McGraw-Hill, 1916). For a summary of early government intervention in the oil sector see John Ise, *United States Oil Policy* (New Haven: Yale University Press, 1928). For a more recent history up to and including the Eisenhower administration see Gerald D. Nash, *United States Oil Policy 1890–1964* (University of Pittsburgh Press, 1968).

[16] Bacon and Hamor, *The American Petroleum Industry*, 242. For annual production figures by state and on a national basis from 1859 to 1914, see 256–7.

The Standard Oil Company was one form of private response to the uncontrolled development of oil.[17] It initially dominated in freight before moving into refining. A 1907 study by the Bureau of Corporations attested to the market power of Standard Oil, finding that along with its affiliates the firm controlled 84.2 percent of refinery runs, and produced 86.5 percent of the national output of kerosene.[18] The same report found that 88.2 percent of petroleum dealers reported purchasing product only from firms affiliated with Standard Oil. This gave the company considerable monopsony power, enabling it to control prices.[19] Oil production, on the other hand, remained relatively unconcentrated, with Standard Oil interests controlling no more than seventeen percent of total oil produced within the United States in 1905, and the organization of production continuing to be made up of large numbers of small

[17] Libecap and Wiggins note the consolidation of production rights is one solution to the common pool problem that plagued US industry, although in this case production rights were controlled through domination of the downstream. Gary D. Libecap and Steven N. Wiggins, "Contractual Responses to the Common Pool: Prorationing of Crude Oil Production," *The American Economic Review* 74 (1984): 87–98. See also Ise, *United States Oil Policy*, 255; McLean and Haigh, *The Growth of Integrated Companies*, 58–62. The relationship between the Standard Oil companies changed over time. The companies formed a trust in 1882, which was dissolved in 1892. The Standard Oil (New Jersey) was then incorporated in 1898. See Department of the Interior, *Report of the Commissioner of Corporations on the Petroleum Industry Vol. 1: Position of the Standard Oil Company in the Petroleum Industry* (Washington, DC: Government Printing Offices, 1907), 2–6.

[18] About seventy-five refiners operated independently from Standard Oil interests, however, their total crude utilization was less than twenty percent of Standard Oil. The Bayonne, NJ, refinery owned by Standard Oil had a larger throughput than all the independent refiners combined. The share of oil refined also varied by field, from 56.9 percent in the small Corsicana field, to 99.1 percent in the mid-continent. Department of the Interior, *Report of the Commissioner of Corporations on the Petroleum Industry Vol. 1*, xvi and 1.

[19] Data from Harold Francis Williamson, *The American Petroleum Industry* (Evanston, IL: Northwestern University Press, 1959), 235; McLean and Haigh, *The Growth of Integrated Companies*, 71. A 1905 House Resolution asked the Secretaries of Commerce and Labor to investigate the structure of the domestic market. These found that "the real source of the Standard's power is not superior efficiency, but unfair and illegitimate practices." Department of the Interior, *Report of the Commissioner of Corporations on the Petroleum Industry Volume II: Prices and Profits* (Washington, DC: Government Printing Offices, 1907), 669. An early but detailed account of the rise Standard Oil is Gilbert Holland Montague, *The Rise and Progress of the Standard Oil Company* (New York: Harper and Brothers, 1903).

operators.[20] Thus the existence of domestic resources initially led to a bifurcation of the industry between few firms with a presence across both refining and production, but concentrated in the former, and many firms focused solely on the production of oil.

Regulation at the state and federal levels played a limited role in shaping the types of firms that emerged. New York introduced legislation to limit waste in 1879, followed by Pennsylvania in 1881, and other states followed suit.[21] The dominant market position of the Standard Oil Company had little to do with these initiatives, however. Most obviously, Standard Oil enjoyed a dominant market position in refining and transport prior to their imposition.[22] An investigation by the Department of the Interior, for example, argued its dominance emerged through the abuse of market power in the form of "railroad discrimination, abuse of control of pipe lines, and unfair methods of selling products."[23] The business practices of the firm led the Department of Justice to open proceedings against the firm, ending with the affirmation of an order to dissolve Standard Oil (New Jersey) in 1911. [24]

Regulating production through the states was also ineffective because of ongoing new finds and the lack of measures to limit interstate trade in oil. The effort to regulate production at the federal level emerged following the discovery of large deposits of oil in California and East Texas, along with stagnant demand. The Federal Oil Conservation Board (FOCB) was created on December 18, 1924, to investigate the state of the industry. As in France and Japan, evidence suggests this push for federal regulation was influenced by the demands of firms. Nordhauser reports, for example, that "the files of the Bureau of Mines and FOCB indicated that every call for conservation by federal authorities was preceded by a corresponding initiative from a major company executive, and was reviewed by organized business operating

[20] There was some variation by field. Department of the Interior, *Report of the Commissioner of Corporations on the Petroleum Industry Vol. 1*, 8–9.

[21] Nash, *United States Oil Policy 1890–1964*, 15.

[22] Montague notes, for example, that the Standard Oil Company controlled over half the country's refining capacity by 1874. Gilbert Holland Montague, "The Rise and Supremacy of the Standard Oil Company," *The Quarterly Journal of Economics* 16 (1902): 281.

[23] Department of the Interior, *Report of the Commissioner of Corporations on the Petroleum Industry Vol. 1*, 22.

[24] Bacon and Hamor, *The American Petroleum Industry*, 260–71.

through the API."[25] The instrument used to regulate the interstate transport of oil was the National Industrial Recovery Act, followed by the Connally Act of 1935 when the former was declared unconstitutional. The Connally Act made it an offense to ignore the quota settings established by the Texas Railroad Commission.[26] Together, these state and federal limits on production were the primary instrument through which business and government oversaw supply in the domestic oil market until the rise of Middle Eastern oil led to the introduction of additional measures in the 1950s. While identifying the precise effects of these regulations on firms is difficult, taken together, it is plausible that they tended to consolidate the characteristics of firms by protecting the interests of those that otherwise would have suffered from falling prices.

Promoting production

If one component of the domestic compact negotiated between firms and the government aimed to regulate the rate of production, firms also sought to enlist the support of government in improving their commercial position in the Middle East and elsewhere, as well as within the national markets of European countries and Japan.[27] In 1919 the Secretary of the Interior proposed that the government should support US firms' access to crude supplies internationally.[28]

[25] Norman E. Nordhauser, "Origins of Federal Oil Regulation in the 1920s" *The Business History Review* 47 (1973): 71.

[26] Williamson, *The American Petroleum Industry*, 548–9. In 1929 an executive order also halted the issuing of new leases for drilling on federal lands, although oil production from federal lands made up only ten percent of national production. Stephen J. Randall, *United States Foreign Oil Policy since World War I: For Profits and Security*, 2nd edn. (Montreal: McGill-Queen's University Press, 2005), 100–4; see also Nash, *United States Oil Policy 1890–1964*, 98–111.

[27] This may have reflected uncertainty about the supply of oil domestically. A 1918 study by the United States Geological Survey, for example, estimated that domestic recoverable reserves would be exhausted by 1928. Despite producing less than five percent of world production, British Petroleum and Royal Dutch Shell acquired more than half the estimated total of non-US reserves by 1919. John A. DeNovo, "The Movement for an Aggressive American Oil Policy Abroad, 1918–1920," *The American Historical Review* 61 (1956): 865.

[28] United States Department of the Interior, *Reports of the Department of Interior 1919 Vol. 1* (Washington DC: Government Printing Office, 1919), 22–4.

In the same year, State Department officials were instructed to support US oil interests operating abroad.[29] One instrument available to promote US firms' interests was the Mineral Leasing Act, passed in 1920. The Act was invoked over the Dutch East Indies, for example, unless the Dutch government gave reciprocal rights to US firms. In 1928 Shell was recognized as an equal under the act, after Standard Oil (New Jersey) subsidiary Koloniale was granted access to the Djambi fields in Central Sumatra.[30] Some evidence suggests the drive to secure reserves internationally for US firms was not only a function of the demand for support from US firms. Fears of oil depletion in the United States were expressed by the US Geological Survey and Bureau of Mines. Indeed, these fears were criticized by industry.[31] In reviewing business–government relations, Gulf Oil Vice-President Charles Hamilton

> recall[ed] very clearly that representatives of industry were called to Washington and told to go out and get [oil] ... We have been encouraged to believe that we were furthering the interests of our country by proceeding to explore the foreign countries for oil.[32]

[29] The Senate passed a resolution on March 10, 1920, calling for the president to identify how US commercial interests were discriminated against in the exploration, development, and production of crude oil internationally. The report, and diplomatic correspondence between the Secretary of State and those of Great Britain in pursuit of US interests, are found in US Department of State, "Message from the president of the United States Transmitting in Response to Senate Resolution No. 149 of February 13, 1924: A Report of the Secretary of State Relative to the Diplomatic Correspondence in Connection with Securing Oil Concessions for American Citizens between this Government and the Governments of Certain Foreign Countries Regarding Oil Concessions in those Countries" (Washington, DC: Government Printing Office, 1924). For an analysis see DeNovo, "The Movement for an Aggressive American Oil Policy Abroad."

[30] Nash argues the diplomatic efforts to support US firms were ineffective in the Middle East and the Asia-Pacific, although more successful in Latin America. See Nash, *United States Oil Policy 1890–1964*, 49–71.

[31] See, for example, US Senate Special Committee Investigating Petroleum Resources, *American Petroleum Interests in Foreign Countries*, 79th Congress, 1st session, 1945, 298–302.

[32] Statement of Charles W. Hamilton, vice-president, Gulf Oil Corporation, on July 27, 1945, *American Petroleum Interests in Foreign Countries*, US Senate Hearings Before a Special Committee Investigating Petroleum Resources, 55. State Department representative Charles Rayner noted

The effects of World War II on oil market governance were transitory, with little impact on the characteristics of firms despite proposals that the government take a more direct role in oil production.[33] In September 1939 executive power was delegated to the Secretary of the Interior, and in December 1942 the Petroleum Administration for War was created by executive order.[34] It held statutory authority for coordinating the production, transportation, and distribution of petroleum.[35] The Petroleum Reserves Corporation was created in 1943 on the grounds that domestic oil reserves were dwindling, and that the government should seek to gain greater leverage over production internationally – specifically Saudi Arabian reserves managed by the California Arabian Standard Oil Company – co-owned by Standard

that the "Department of State has ... taken the position that the public interest of the United States requires ... large-scale expansion of holdings in foreign oil reserves by United States Nationals. It has, therefore, actively supported the efforts of United States petroleum interests to secure and to consolidate concessions abroad." Statement of Charles Rayner, Petroleum Adviser, Department of State, on February 10, 1944, to Senate Committee Investigating the National Defense Program, cited in *American Petroleum Interests in Foreign Countries*, 1. Rayner notes that the State Department assigned thirteen or fourteen oil advisors to relevant embassies to support firm activities internationally.

33 Stoff recounts the effects of the war on US firms' business interests. Michael B. Stoff, "The Anglo-American Oil Agreement and the Wartime Search for Foreign Oil Policy," *The Business History Review* 55 (1981): 59–74. Randall argues government was not captured by industry interests during the war. Stephen J. Randall, "Harold Ickes and United States Foreign Petroleum Policy Planning, 1939–1945," *The Business History Review* 57 (1983): 367–87. Keohane reaches a similar conclusion: "State officials had their own purposes and were not merely the tools of private interests, but they were constrained step by step, by the political power of the oil companies." Robert O. Keohane, "State Power and Industry Influence: American Foreign Oil Policy in the 1940s," *International Organization* 36 (1982): 166.

34 Nash, *United States Oil Policy 1890–1964*, 159–60.

35 In a review that is otherwise sympathetic to the independence of government from industry interests, Randall concludes that "business-government links were significant in the PAW, and as strong a personality as Ickes was, he was heavily reliant on his deputy administrator, Ralph Davies, a vice-president and director of Standard Oil of California (Socal), who clearly determined operational decisions even if broader policy objectives were set by Ickes." Randall, "Harold Ickes and United States Foreign Petroleum Policy Planning," 369. A summary of US firms' international holdings can be found in *American Petroleum Interests in Foreign Countries*.

Oil (California) and Texaco.[36] Texaco baulked at transferring its share in CASOC, however; industry resisted the idea of direct government ownership of assets in any stage of the industry.[37] The consensus view of firms was summarized in the National Oil Policy issued by the National Oil Policy Committee within the Petroleum Industry War Council, which noted that the status quo in which private industry remained autonomous in its international operations, but could call on diplomatic support when needed, was preferable.[38]

Loss of competitiveness and rise of protectionism

The demand from firms for government intervention in support of their international position is a common element across all three cases. As crude production in the Middle East grew, a new problem emerged in the politics of domestic oil market governance in the US that was necessarily absent in the French and Japanese cases given their lack of domestic resources: rising crude oil imports. Imports of crude oil into the United States grew from 5.8 percent of total supply at the end of the war to 9.9 percent by the end of 1949. The marginal growth came largely from Saudi Arabia and Kuwait, although the biggest share of imports continued to be constituted by oil from Venezuela and the Dutch East Indies.[39]

The effect of these imports on firms depended on their international position. Rising imports hurt the position of domestically focused

[36] "Letter to the president, June 26, 1943," in US Senate Committee on Foreign Relations, *A Documentary History of the Petroleum Reserves Corporation 1943–1944* (Washington, DC: US Government Printing Office, 1974), 9–12.

[37] "Minutes of a Special Meeting of the Directors of the Petroleum Reserves Corporation, September 13, 1943," in Senate Committee on Foreign Relations, *A Documentary History of the Petroleum Reserves Corporation 1943–1944*, 27; "Minutes of a Special Meeting of the Directors of the Petroleum Reserves Corporation, September 28, 1943," in Senate Committee on Foreign Relations, *A Documentary History of the Petroleum Reserves Corporation 1943–1944*, 31.

[38] "A National Oil Policy for the United States: A Preliminary Report of the National Oil Policy Committee of the Petroleum Industry War Council," February 28, 1944, in Senate Committee on Foreign Relations, *A Documentary History of the Petroleum Reserves Corporation 1943–1944*, 69–80.

[39] Import figures are drawn from US Select Committee on Small Business, "Effects of Foreign Oil Imports on Independent Domestic Producers," Hearings before the Select Committee on Small Business, House of Representatives Parts 1–3, May 25, 1949–May 2, 1950, 974–6.

firms, and the IPAA responded by arguing for import barriers: "increasing imports mean less income to the domestic industry, less drilling, less reserves found, less productive capacity. In short, less security as to oil."[40] Congressional members from oil-producing districts were sympathetic to this view. The representative from New Mexico, for example, noted that crude oil output represented the largest source of income for the state, and that oil sector employment was threatened by imports. Illinois representative Charles W. Vursell summarized his position by linking domestic industry with national security:

> My principal interest in introducing and supporting this legislation is my interest in national defense. My secondary interest is in the fact that my State of Illinois is the sixth State in the Union in the production of oil, and I may say to the committee that up to the present time Illinois has produced 1,600 million barrels of oil.[41]

[40] Statement Of H.B. Fell, president, Simpson-Fell Oil Co., Ardmore, OK, and executive vice-president, Independent Petroleum Association, on April 27–May 19, 1953, to the US House of Representatives Hearings Before the Committee on Ways and Means on the Trade Agreements Extension Act of 1953, 83rd Congress, 1st session, 1953, 1010. The vice-president for Sun Oil noted that "You can't tell the domestic producer that imports at present are not interfering with exploration and development. He knows better. He knows allowables are cut back. He knows hundreds of drilling rigs are standing idle. He knows the current rate of drilling is below that necessary to achieve the requirements set to meet defense goals. He finds his profits down and new funds for drilling hard to obtain. And you just can't make him believe there is nothing wrong. When he looks around to see what is wrong, the thing that stares him in the face is the fact that imports have been mounting at the same time that allowables in Texas have been cut back by as much as 384,000 barrels daily since last December. He's smart enough to put the pieces of that puzzle together and come up with the answer: imports are supplanting his production and taking away his market." See 979. Also see Statement of Russell B. Brown, General Counsel, Independent Petroleum Association of America, on April 27–May 19, 1953, to the US House of Representatives Hearings Before the Committee on Ways and Means on the Trade Agreements Extension Act of 1953, 83rd Congress, 1st session, 1953, 927–40.

[41] Statement of Honorable Antonio Fernandez, Representative in Congress from the State of New Mexico, on April 27–May 19, 1953, to the US House of Representatives Hearings Before the Committee on Ways and Means on the Trade Agreements Extension Act of 1953, 83rd Congress, 1st session, 1953, 224–5; Statement of Honorable Charles W. Vursell, Representative in Congress from the State of Illinois, on April 27–May 19, 1953, to the US House of Representatives Hearings Before the Committee on Ways and Means on the Trade Agreements Extension Act of 1953, 83rd Congress, 1st session, 1953, 920.

The majors, on the other hand, benefitted from the rise in oil from the Middle East so long as they controlled the rate of production. Standard Oil (New Jersey) began to produce more oil from fields in the Middle East than in the United States by 1958, and by 1960 Middle Eastern oil reached a quarter of total production, while US production stood at one in five barrels.[42] Middle East and Libyan oil reached forty-five percent of total firm production in 1972.[43] This shift was reflected in the position they took towards limiting oil imports. In a letter to shareholders, Standard Oil (New Jersey)'s managers argued:

> The Company has consistently opposed legislation to restrict imports. We recognize that a healthy domestic oil industry is vital ... Jersey believes, however, that artificial restrictions such as quotas on oil imports would react to the industry's own detriment and would be contrary to the best interests of both consumers and the national welfare.[44]

Other firms with a significant international position argued likewise. Standard Oil of California, for example, also opposed the imposition of quotas on oil. Chairman of the Board R.G. Follis argued that the firm was "opposed to legislation of the type in the proposed bill, because it destroys the flexibility so necessary in an industry such as petroleum in meeting rapidly changing conditions in the markets." Gulf Oil agreed:

> placing of arbitrary restrictions upon the importation of oil into the United States is not only unnecessary but would be very unwise from the standpoint of our growing oil needs in this country, our defense position in the world, and our trade and diplomatic relations with other countries.[45]

[42] Figures calculated from gross production of crude oil and natural gas liquids. Standard Oil (New Jersey), *Annual Report Standard Oil Company 1960 (New Jersey)* (New York: Standard Oil [New Jersey], 1960), 14.

[43] Exxon Corporation, *Exxon Corporation 1972 Annual Report* (New York: Exxon Corporation, 1972), 36.

[44] Standard Oil (New Jersey), *Annual Report Standard Oil Company 1953 (New Jersey)* (New York: Standard Oil [New Jersey], 1953), 6.

[45] Statement of chairman of Board of Standard Oil California, on April 27–May 19, 1953, to the US House of Representatives Hearings Before the Committee on Ways and Means on the Trade Agreements Extension Act of 1953, 83rd Congress, 1st session, 1953, 1284; Statement of Sidney A. Swensrud, chairman of the Board, Gulf Oil Corp., on April 27–May 19, 1953, to the US House of Representatives Hearings Before the Committee on Ways and Means on

Congress took up the question of how to respond to the rise in imports by examining whether to amend the Trade Agreements Extension Act of 1955 to give the president the authority to restrict imports for reasons of national security.[46] The IPAA petitioned the Office of Defense Management (ODM) to restrict oil imports in the name of national security once this amendment was passed. On July 29, 1957, President Eisenhower called for voluntary restraints on imports. In February 1958 sanctions for non-compliance were increased.[47] Presidential Proclamation 3279 then established a mandatory system of import controls, with import licenses issued by the Secretary of the Interior.[48] The quota program remained in place for fourteen years.[49]

Most fundamentally, the barriers to trade were abolished because policymakers could not bear the costs of rising product prices implied by the peaking of US production.[50] Peak oil did not mean an immediate decline in US production, but the quota system assumed domestic production could meet demand growth without causing significant

the Trade Agreements Extension Act of 1953, 83rd Congress, 1st session, 1953, 1240.

[46] For a summary see William B. Kelly, Jr., "The 'Expanded' Trade-Agreements Escape Clause 1955–61," *Journal of Political Economy* 70 (1962): 37–63.

[47] Richard E. Mooney, "New Curb Studied on Imports of Oil," *New York Times*, February 9, 1958: 29; Richard E. Mooney, "Eisenhower Sets Oil Import Curbs," *New York Times*, March 28, 1958: 1.

[48] For details see R.O. Kellam, "Regulation of Oil Imports," *Duke Law Journal* (1961): 175–215. For a summary of the structure and operation of the program see Richard H. K. Vietor, *Energy Policy in America since 1945: A Study of Business–Government Relations* (Cambridge University Press, 1984), 119–45.

[49] "White House, Statement by the president, March 10, 1959," cited in Douglas R. Bohi and Milton Russell, *Limiting Oil Imports: An Economic History and Analysis* (Baltimore, MD: Johns Hopkins University Press, 1978), 67. The program was reviewed and amended numerous times over this period. The amendments authorized through presidential proclamation are reprinted in Bohi and Russell, *Limiting Oil Imports*, 91–8. While the goal of this section is to try to understand the origins of firm characteristics, rather than trace the reasons for changes in oil market governance in the US market, it is worth noting that the different positions firms took towards the rise of imports match expectations. The divergence in firm characteristics complicated relations between firms, setting the majors against smaller independents focused on the domestic market. Congressional testimony suggests that the larger and more internationally diversified firms opposed the imposition of barriers to trade in crude oil, while smaller and domestically focused firms, along with the industry association that represented them, supported this change.

[50] The politics of its abolition are discussed in Bohi and Russell, *Limiting Oil Imports*, 193–201.

price increases. The plateau in US oil production undermined this assumption.[51] President Nixon established a Cabinet Task Force on Oil Imports to clarify the performance of the system. It concluded that supply cutoffs mattered because of their potential effect on defense production and the civilian economy, but that the quota system was poorly designed. Quotas were abolished in 1973. Secretary of the Interior Morton described the reason for this change:

> The original import program was adopted at a time when the country had excessive oil producing capacity, the world was flooded with low cost foreign oil and quantitative import quotas were needed to maintain domestic producing capacity in the interest of national security. Today, the situation has changed radically. We are not capable of producing as much oil as we are consuming; thus, we must increase imports to avert short term fuel shortages. Yet, we must concurrently take steps to limit or reduce our long term reliance on oil imports by encouraging domestic petroleum development and the construction of more refining capacity to process it.[52]

A second change was the abolition of price controls. In 1971 the Nixon White House froze prices of petroleum products, in addition to wages and prices for other goods. They remained for both oil and oil products under the Emergency Petroleum Allocation Act (EPAA) of 1973.[53] The EPAA enabled a two-tiered pricing system. Prices were fixed for "old" oil, while prices for oil produced from new wells – or incremental increases above April 1972 production – were allowed to

[51] Exemptions to the import quota program were common, which undermined the operation of the quota system. For a summary and industry responses see US Senate Committee on the Judiciary, "Government Intervention in the Market Mechanism: The Petroleum Industry Part Two – Industry Views," Hearings Before the Subcommittee on Antitrust and Monopoly of the Committee on the Judiciary, United States Senate, May 20–9, 1969; "Government Intervention in the Market Mechanism: The Petroleum Industry Part Three – Complainants' Views," Hearings Before the Subcommittee on Antitrust and Monopoly of the Committee on the Judiciary, United States Senate, July 22–31, 1969.

[52] Rogers C.B. Morton, "The Nixon Administration Energy Policy," *Annals of the American Academy of Political and Social Science* 410 (1973): 65–74; Thomas O'Toole, "Energy Plan Ends Quotas on Oil Import," *Washington Post*, April 19, 1973: A1.

[53] Price ceilings for most products were removed over the 1976 to 1979 period. See Joseph P. Kalt, "The Creation, Growth, and Entrenchment of Special Interests in Oil Price Policy," *Harvard Institution of Economic Research Discussion Paper Series* 939 (1982): 1–33.

float. President Ford then began the process of decontrolling prices on "old" oil in order to cut demand and promote production.[54] This was completed under the Reagan administration. Domestic oil-producing firms and congressional members from oil-producing states supported decontrolling prices.[55] They were joined by the American Petroleum Institute (API), which agreed that the best way to go about increasing the production of crude within the United States was through price decontrol, thereby increasing the incentives for firms to explore and produce within the United States.[56]

Oil market governance and firm characteristics

The case of the United States offers a useful extension to the French and Japanese cases examined in the previous two chapters. One reason for this is that it enables us to probe the relative importance of industrial compacts, and initial resource endowments, in shaping the characteristics of firms. What does the brief review above suggest?

It is first worth noting that, despite this difference in initial conditions, there are some elements that are common across the French, Japanese, and US cases. Most simply, a component of the strategies adopted by firms in all three cases focused on using the resources of government to improve their commercial position. Indeed, in all three

[54] See White House, "A Transcript of the State of the Union Message to Congress by President Ford," *New York Times*, January 16, 1975: 24; William D. Smith, "Most of Oil Industry Backing Ford's Plans," *New York Times*, May 29, 1975: 47.

[55] The IPAA argued that price decontrol would lead to two million barrels per day of extra production within a decade. Staff reporter, "US Controls Phaseout Seen Adding 2 Million b/d," *Oil & Gas Journal*, February 26, 1979: 41. A sample of the positions of actors towards decontrol is in US House of Representatives, "President's Decontrol Policies," Hearings Before the Subcommittee on Energy and Power of the Committee on Interstate and Foreign Commerce, United States House of Representatives, 94th Congress First Session, July 11, 14, 28, and September 8, 9, 1975.

[56] In a 1979 interview following the announcement to implement price decontrol, the head of the API, Charles DiBona, argued price decontrol would allow oil firms to invest more in domestic exploration and production. He noted API estimates showed price decontrol would lead to an extra 1.5–2 million barrels a day of crude production by the middle of the 1980s, and praised President Carter for making the decision to implement the policy. Staff reporter, "DiBona: Decontrol Could Buoy US Oil," *Oil & Gas Journal*, April 16, 1979: 37.

cases, one outcome of the process of negotiation between firms and government is the imposition of barriers to trade. On the other hand, the existence of domestic resources in the United States introduced an element absent in the French and Japanese cases: promoting the interests of firms by subsidizing the domestic production of oil. This made the strategies adopted by firms more varied than the cases of both France and Japan, both because of the need to manage the problem of overproduction, and because it gives policymakers an incentive to support firms that produced within the United States.

How did the existence of a substantial endowment of oil in the United States, and the industrial compacts negotiated between business and government, affect the characteristics of firms? The brief review above suggests the existence of substantial reserves of oil within national borders – something both France and Japan lacked – was crucial in shaping the characteristics of firms. This can most clearly be substantiated by the fact that large and vertically integrated US firms emerged prior to the emergence of regulations governing oil at the federal and state levels. While these firms at times called on the instruments of the state to improve their competitive position internationally, it is not clear how effective this support was. The provision of diplomatic support also appeared to follow, rather than precede, the development of these firms.[57]

Just as the industrial compacts negotiated between business and government in the cases of France and Japan led to different types of firms emerging across these countries, the existence of domestic resources in the case of the United States similarly led to the emergence of different types of firms. In addition to the large, diversified firms that came to dominate oil internationally, the existence of abundant domestic resources of oil also enabled the emergence of a large number of smaller, domestically focused producers. These firms remained salient politically, and successfully sought protection as oil produced by the majors in the Middle East began to dominate the supply of incremental increases in oil consumption in the 1950s. The relative

[57] A US Senate review of diplomatic support for US firms argued the government played an important role during initial negotiations with the British over the Open Door in the Middle East, but that in the 1930s "government intervention in [sic] behalf of industry interests ... has been confined almost exclusively to consultation service after an issue has evolved." See *American Petroleum Interests in Foreign Countries*, 313.

abundance of oil in the United States therefore led not to the emergence of diversified firms operating at scale, but rather to the bifurcation of the industry at the production stage between the progeny of the Standard Oil Company and a large number of small firms producing within the domestic market alone.

The review also suggests that the industrial compact negotiated between business and government in the United States played a role in shaping firm characteristics. Rather than decisively determining the types of firms that emerged, however, it tended to consolidate these differences in firm characteristics. Federal regulation of production from the 1930s established a floor on prices, protecting less competitive firms from insolvency. Most importantly, as imports increased as a ratio of total domestic consumption, government acceded to the demands of smaller US producers by erecting barriers to trade.

Estimating the effects of the benefits that accrued to domestic firms through these measures is difficult because it requires making assumptions about oil prices in the absence of the trade restraints. One estimate nevertheless found that domestic production could have halved under conditions of open trade.[58] The IPAA also argued that without import controls the number of firms bankrupted would have been up to four times greater than the 3,000 firms that went out of business between 1956 and 1968.[59] While the magnitude of the effect is thus difficult to measure, it suggests that support for these firms had a substantial effect on consolidating this industry structure, just as protection in the refining sector consolidated the small and poorly diversified structure of firms in Japan. This also served to retain smaller producers as an important interest group in US oil politics. Thus, government policies tended to "maintain the basic thrust of the initial factor endowment," in this case the emergence of two main types of firms

[58] Bohi and Russell find that domestic crude production would have been 3.5 billion barrels in 1960, compared to an actual figure of 2.9 billion barrels. This difference would have grown to 4.1 billion barrels in 1970, compared to 2.2 billion barrels under no limits on trade. Bohi and Russell, *Limiting Oil Imports*, 277–85.

[59] Statement of Harold McClure, Jr., president of the Independent Petroleum Association of America, on May 20, 1969, to the US Senate Hearings Before the Subcommittee on Antitrust and Monopoly of the Committee on the Judiciary, "Government Intervention in the Market Mechanism: The Petroleum Industry Part Two – Industry Views," 91st Congress, 1st session, 1969, 618.

producing oil in the United States.[60] As I suggest below, smaller producers remained a powerful political force in shaping oil market governance in the United States following the transformation of the international oil market.

Explaining changes in oil market governance since 1980

The second reason the US case is a useful extension of the previous two chapters is because it enables us to examine the effect of firm characteristics on the demands they made on government unencumbered by cross-national differences in political institutions. If the argument advanced in Chapter 3 is correct, we should see larger and more diversified firms more capable of taking advantage of the diversification of the international oil market and, able to manage oil price volatility, less interested in support from the state. Smaller and less diversified firms, on the other hand, should be more willing to seek support from government.

In this section I move on to examine changes in oil market governance in the United States since 1980. An important difference between the United States case and those of France and Japan is that emergence of a more liquid, but volatile, international oil market coincided with the end of growth in domestic oil production within the US itself until the unconventional oil revolution decades later. In this section I consider how the shift in opportunities and risks associated with the emergence of an international market for oil affected the strategies of US firms, and how this was influenced by their characteristics. In order to explore this question I conduct two short case studies. The first focuses on the period of falling crude oil prices of the 1980s. I find this led to incremental adjustments to existing laws governing the domestic market: the date for completing price decontrol was brought forward, leasing of federal lands and the Outer Continental Shelf were loosened, and long-established subsidies for domestic industry remained in place. Trade barriers, on the other hand, were not reintroduced despite demands from some members of Congress and a number of firms.

[60] Stanley L. Engerman and Kenneth L. Sokoloff, "Factor Endowments, Institutions, and Differential Paths of Growth Among New World Economies: A View from Economic Historians of the United States," *NBER Working Paper Series on Historical Factors in Long Run Growth* 66 (1994): 35.

The second case study covers the period of rising prices beginning in the 2000s. The administration of George W. Bush entered into government declaring its intention to implement a national energy policy by making changes to market governance in electricity and natural gas, in addition to oil. The political salience of oil for consumers also increased along with the rise in oil product prices. Passing legislation nevertheless proved challenging given the large number of environmental and other trade-offs legislators forced one another to make. The events of September 11 also dominated policymaking in 2001 and 2002, slowing deliberation over energy policy. Congress nevertheless sent a bill to the president during his second term, which once again included subsidies designed to enhance the competitiveness of US oil.

In the previous two chapters on France and Japan I examined how firms responded to the change in oil market structure, and how this was influenced by their characteristics. The US case enables us to examine how the characteristics of firms affected their demands under the same set of political institutions. In both of the cases considered below negotiations between firms, other interest groups, and government over how to respond to oil price volatility were fraught, as in the French and Japanese cases. Disagreements also emerged between Congress and the executive branch. In addition, industry itself was divided over the merits of erecting barriers to trade, although it was united on the merits of increasing opportunities within the United States using subsidies and other incentives.

Congressional testimony confirms the similarities in the strategies that firms adopted in response to the changed market conditions of the 1980s across all three cases, depending on their characteristics. In the United States the most difficult period for smaller and less diversified firms was the collapse in oil and oil product prices in the 1980s. This harmed profitability, just as it hurt both the Japanese oil firms operating in exploration and production and the Japanese refiners. The response of some of these firms was to push for government support in order to improve their competitive position against imports. This was met, as in Japan, with opposition from manufacturers and other consumers of final products, who feared that measures to increase the price of oil and oil products domestically would reduce their competitiveness. In the case of Japan, this opposition was an important reason for the gradual abolition of barriers to petroleum product imports. In the United States, it formed part of a powerful coalition against the reimposition of barriers to imports. Proposals to reintroduce barriers to both oil imports and imports of oil products faced strong and unified opposition in Congress from consumers, as well as from congressional members from districts and states that had little oil production and therefore did not have a strong incentive to support the interests of the industry. The executive branch also opposed amendments that were likely to lead to the introduction of a tariff on oil. More diversified firms, on the other hand, responded to the fall in prices by reducing capacity, diversifying operations further, and at times merging with other firms, as in France. They also opposed the imposition of barriers to imports, just as French firms were disinterested in maintaining barriers to imports. In contrast to these divisions over the question of import tariffs, testimony shows firms, and the industry associations representing them, were united in supporting improving access to supplies of crude on federal lands, as well as retaining and increasing subsidies, options that were unavailable to both French and Japanese firms because of the lack of domestic resources.

Case study 1 – falling prices and oil market governance

The 1980s were characterized by a period of falling oil prices. The cost of acquiring crude oil for refiners, for example, fell from a little under twenty-five dollars a barrel at the start of 1980, to just under fourteen dollars at the end of 1986 in nominal terms. Product prices also fell. The price of a gallon of unleaded gasoline, for example, dropped from one dollar thirteen cents to eighty-two cents over the same period.[61] Given this, it is unsurprising that congressional testimony suggests there was little pressure from consumers to reduce the costs of petroleum products, although consumer and industry groups testified against the reintroduction of import barriers.

Data shows the fall in prices also led to a sharp fall in domestic oil production, and this particularly hurt the performance of firms focused on producing oil in the United States. What effect did these changes have on the strategies they adopted, and the willingness of policymakers to

[61] Data from US Energy Information Agency. Available from www.eia.gov/petroleum.

support them? Evidence drawn from congressional testimony, annual reports, and other sources, suggests that falling oil prices led independents, as well as some members of Congress, to support measures designed to shore up the position of domestic producers. These measures included retaining and increasing the incentives for domestic production.

Testimony also suggests that firms were not united in the demands they made on government, however. Instead, the US majors supported increasing incentives to oil development in the United States in unison with domestically focused firms, but opposed the reimposition of barriers to trade. While there are no votes that focus specifically on the issue of reintroducing oil imports that allow us to clearly identify the positions of members of Congress on the issue, testimony does suggest the split in the strategies of firms was mirrored among policymakers: members from oil-producing districts tended to support trade barriers in order to protect domestic firms, while others were not interested in boosting production through the introduction of tariffs. There was also a division between the executive branch and some congressional members over how to respond to falling oil prices.

The executive branch also opposed the reintroduction of trade barriers. Prior to entering office the Reagan administration signaled that it opposed the more interventionist position adopted by the Carter administration, arguing for "removing government obstacles to energy production."[62] This position had the good fortune of being unlikely to be costly politically, in addition to being ideologically consistent with other initiatives, given that falling prices meant public demand for intervention was less than in the previous decade. The first oil-related decision made by the White House was to exempt crude and refined petroleum products from price and allocation controls. This required no congressional action, and brought forward by eight months the date determined by the Carter administration.[63] Secretary of Energy Edwards stated that moving decontrol forward was designed to "stimulate greater efficiency and productivity in the US economy."[64] The White

62 W.A. Bachman, "Where the Presidential Hopefuls Stand on Energy," *Oil & Gas Journal*, March 31, 1980: 51.

63 Exceptions already granted meant that the decision was relevant for fifteen percent of total US crude production. Staff reporter, "Reagan: End Controls on Crude, Products," *Oil & Gas Journal*, February 2, 1981: 25.

64 Staff reporter, "Edwards: Allocation not Favored in Crude Supply Pinch," *Oil & Gas Journal*, February 23, 1981: 41.

House also argued that price decontrol served to decrease dependency on foreign oil by stimulating domestic production, and the measure was noted by the White House as one of its early achievements.[65]

A second proposal focused on promoting production through easing the leasing of federal lands and the OCS. According to a 1980 assessment by the API, the federal government owned 775 million acres onshore and 528 million acres on the OCS, accounting for thirty-seven percent of undiscovered oil resources, but were responsible for sixteen percent of national production. It further estimated that eighty percent of recoverable reserves of oil shale and ninety-five percent of tar sands were located in these areas.[66] In 1981 the White House announced that it would increase the number and acreage of leases made available under the OCS Lands Act, and expedite the regulatory procedures covering leasing of onshore federal lands in order to make more land available to firms interested in exploring for oil.[67]

Changes were also made to the fiscal regime governing oil production. The Windfall Profits Tax enacted under the Carter administration was cut during the first term. Initial changes were made through the Economic Recovery Act of 1981, which reduced the tax on revenues secured by firms operating in the petroleum sector as a result of price rises under gradual decontrol. The tax was fully repealed as part of the 1988 Omnibus Trade and Competitiveness Act, while subsidies for domestic production were kept in place. In its last months the White House concurred with a finding that oil imports represented a threat to national security under Section 232 of the Trade Expansion Act, but refused to implement any measures in response.

How can we explain this outcome? While the existence of domestic oil production and the larger number of veto-players in US decision-making make it difficult to isolate the relationship between outcomes and the relative importance of the positions adopted by each of these

[65] John T. Woolley and Gerhard Peters, *The American Presidency Project* [online]. Santa Barbara, CA: University of California (hosted), Gerhard Peters (database). Available from www.presidency.ucsb.edu/ws/?pid=43427.

[66] Staff reporter, "Outlook Cloudy for More Public Land Energy Work," *Oil & Gas Journal*, November 10, 1980: 139.

[67] John H. Jennrich, "US Moves to Speed Offshore Leasing," *Oil & Gas Journal*, April 20, 1981: 42. This did not require the passing of legislation, as changes were covered by the Mineral Leasing Act and the Lands Act, both administered by the Department of the Interior.

actors, it is clear that the strategies adopted by firms and policymakers are consistent with the argument developed in previous chapters. As a result this group, that opposed the reimposition of tariffs on oil and oil products, represented a powerful constituency of firms, consumers, congressional members, and the executive branch. The coalition opposed to subsidies, on the other hand, was weaker, which is consistent with an outcome in which trade barriers were rejected but other forms of support were retained. As with the French and Japanese cases, I begin by discussing firm strategies, before moving on to outline the positions taken by policymakers.

Stable but divided firm strategies

Oil price volatility in the 1980s had a devastating effect on domestic exploration and drilling. Drilling activity fell by eighty-five percent from its 1981 peak. Data collected by the API and IPAA showed that revenues for domestic producers halved by 1986, while capital expenditures on domestic exploration and production fell by twenty-five percent. Employment dropped by 126,000, or twenty-two percent of total industry employees, in the first seven months of 1986. Falling prices and overcapacity also had an impact on firm revenues and profits. Even for the leading firms, revenues fell by eighteen percent over 1985–6, and return on equity was 9.7 percent for these firms, compared to 14.4 percent for the ninety largest non-oil industrial companies.[68] The fall in prices hurt both diversified and non-diversified firms, but the performance of firms focused on oil production alone, according to API President Charles DiBona, was "much worse."[69] In a 1986 testimony to Congress, the chairman of the IPAA, which represents smaller and less diversified firms, noted that "since 1982, literally thousands of economic entities have been liquidated, including independent producers, drilling contractors,

[68] See statements by API and IPAA presidents to the Senate Committee on Energy and Natural Resources, September 18, 1986. US Senate Committee on Energy and Natural Resources, "Domestic and International Petroleum Situation and the Implications of Fees on Imported Oil," Hearing Before the Committee on Energy and Natural Resources, US Senate, 99th Congress, 2nd session, 1986.

[69] Statement of Charles J. DiBona, president of American Petroleum Institute, on September 18, 1986, to the Hearing Before the Committee on Energy and Natural Resources, US Senate, 99th Congress, 2nd session, 122 and 125–33.

supply and equipment firms, and financial institutions."[70] The API estimated that, at fifteen dollars a barrel, domestic production would fall further, from 8.9 million barrels a day in 1985 to 6.2 million barrels by 1991.[71]

What strategies did firms adopt in response to falling prices, including towards government? Both the US majors and smaller and less diversified independents suffered in the falling price environment. The chairman of Chevron, George M. Keller, noted in his letter to shareholders, for example, that the new price environment represented a "dramatic change in the economics of the petroleum industry that [had] permanently altered the course of our company's future."[72] Evidence suggests that despite this, the US majors, in common with CFP and Elf Aquitaine, were in a better position to manage the fall in oil prices. One response was to pursue growth through merger. Chevron, for example, completed a 13.3 billion dollar deal to acquire Gulf Oil in 1984, while changing the company name from Standard Oil of California to Chevron.[73] The merger almost doubled the proved net reserves owned by the firm, which drilled in twenty-eight countries outside the United States, in addition to wells offshore of Texas, Louisiana, and other areas in the OCS. Although the firm invested in exploration and production opportunities in Alaska and the Gulf of Mexico, it also pursued the diversification of suppliers outside the United States, including in Africa and Asia, as well as increasing the amount of oil purchased on the market, in order to gain "a geographic balance, which gives a measure of insurance

[70] Statement of Raymond H. Hefner, Jr., chairman, Independent Petroleum Association of America, on March 14, 1986, to the US Senate Committee on Energy and Natural Resources, "Domestic and International Petroleum Situation and the Implications of Fees on Imported Oil," Senate Hearings on Petroleum Industry Issues, 99th Congress, 2nd session, March 14, 1986, 197.

[71] Statement of Charles DiBona, president of American Petroleum Institute, to Senate Committee on Energy and Natural Resources, 1986, 131–7.

[72] George M. Keller, "To Our Stockholders," *Chevron Corporation 1984 Annual Report* (San Ramon, CA: Chevron Corporation, 1984), 2.

[73] The chairman noted in his remarks to shareholders that this represented the best response to the falling price environment, in addition to increasing production from established wells, and implementing cost-cutting measures: "New realities have to be faced squarely – including crude oil surpluses, reduced petroleum consumption, unstable prices and sometimes narrow or non-existent profit margins in manufacturing and marketing operations." Keller, *Chevron Corporation 1984 Annual Report*, 3

should a political upheaval cause a supply disruption in one part of the world."[74]

A review of the annual reports of Exxon, Mobil, Texaco, and others, shows they adopted similar strategies. Exxon sought to improve performance by growing reserves and increasing production from existing wells, as well as by cutting refining capacity to reduce costs and adjust to falling demand for oil products. Mobil (formerly Standard Oil of New York) acquired the Superior Oil Company, which was focused on oil production within the United States, for $5.7 billion in 1984, and also sought to further diversify its holdings internationally through investments in the North Sea. Mobil's proved reserves of crude stood at almost six billion barrels worldwide, with 2.5 billion barrels in the US and Canada.[75] This shift towards diversification internationally was noted in a survey of the industry by the US Office of Technology Assessment, which found that the expansion of opportunities led firms to increase international investments.[76]

Given this, firms were split on the demands they made on government. On the one hand, a consensus existed on the benefits of tax policies and subsidies that promoted domestic oil production. The API, with a broad membership base that included both majors and independents, lobbied for the abolition of the Windfall Profits Tax, arguing that US production was 800,000 barrels per day lower because of the tax.[77] This argument was supported by the IPAA, which is unsurprising given that the tax was levied on oil produced within the United States, thus making it less competitive relative to imports.

[74] Chevron Corporation, *Chevron Corporation 1984 Annual Report*, 5–15.

[75] C.C. Garvin, Jr. and H.C. Kauffmann, "To the Shareholders," in *Exxon Corporation Annual Report 1984* (New York: Exxon Corporation, 1984), 2–3. Rawleigh Warner, Jr. and Allen E. Murray, "To the Shareholders," *Mobil Annual Report 1984* (New York: Mobil Corporation), 2–3. Texaco, perhaps reflecting its production profile that saw fifty percent of oil and gas production within the United States, also supported the federal government establishing a floor for domestic oil prices in order to protect domestic production. See John K. McKinley, "To the Stockholders," *Texaco Inc. Annual Report 1984* (White Plains, NY: Texaco Incorporated, 1984), 3.

[76] Office of Technology Assessment, *US Oil Production: The Effect of Low Oil Prices – Special Report*, OTA-E-348 (Washington, DC: US Government Printing Office, September 1987), 15–16.

[77] Statement of Charles J. DiBona, president of American Petroleum Institute, on June 5, 1987, to the US Senate Hearing Before the Subcommittee on Energy and Agricultural Taxation of the Committee on Finance, US Senate, 100th Congress, 1st session, 1987, 172. The API also argued that any decision to

Firm and industry representatives also agreed on the need to retain subsidies. The API argued that increasing the tax burden would "reduce future domestic production, increase reliance on imported oil and play into the hands of OPEC."[78] The president of the IPAA, Lloyd N. Unsell, argued that the industry was going through a "devastating economic shakedown," and that the federal government should introduce measures to improve the performance of the domestic industry.[79] Raymond H. Hefner, representing the IPAA and forty-three state-level independent oil and gas associations, argued that an "aggressive new energy and tax policy is needed that will ... encourage the exploration and development of new domestic oil and gas reserves ... stop the dramatic loss of stripper and marginal production; and ... maximize existing reserves through enhanced recovery."[80] The IPAA focused in particular on intangible drilling costs, arguing these costs represented forty percent of total costs in exploration and drilling, and sixty-five percent of annual expenditures.[81] Regional associations such as the Petroleum Association of Wyoming, which represented firms responsible for ninety percent of the production and wells drilled in the Rocky Mountain States, similarly argued that tax incentives were "an absolute necessity."[82]

increase taxes should focus on broad consumption-based taxes rather than more narrowly based taxes that focused on energy.

78 Statement of Charles J. DiBona, president, American Petroleum Institute, on June 9, 1982, to the US Senate Subcommittee on Energy and Agricultural Taxation of the Senate Committee on Finance, "Energy Tax Options," 97th Congress, 2nd session, 1982, 50–6.

79 Statement of Lloyd No Unsell, president of Independent Petroleum Association of America, on September 18, 1996, to the Hearing Before the Committee on Energy and Natural Resources, US Senate, 99th Congress, 2nd 1986: 2.

80 Statement of Raymond H. Hefner, Jr., chairman, Independent Petroleum Association of America, on March 14, 1986, to the Hearings on Petroleum Industry Issues, "Domestic and International Petroleum Situation and the Implications of Fees on Imported Oil," Senate Committee on Energy and Natural Resources, 193.

81 These benefits were crucial, given many smaller firms depended on earnings to finance future investment. "Independent Petroleum Producers of America: Statement on Alternative Minimum Tax Applied to Intangible Drilling Costs," on June 9, 1982, to the US Senate Subcommittee on Energy and Agricultural Taxation of the Senate Committee on Finance, "Energy Tax Options," 97th Congress, 2nd session, 1982, 80–5. On the importance of intangible drilling costs, John Northington, Special Assistant to Assistant Secretary of the Interior for Land and Minerals, Department of the Interior (1998–2000), interview with author, Washington, DC, June 3, 2011.

82 Statement of Richard T. Robitaille, Executive Director, Petroleum Association of Wyoming, on June 5, 1987, to the Hearing Before the Subcommittee on

Industry also united in favor of increasing opportunities for drilling on federal lands. Industry associations opposed legislation proposing to remove federal lands from leasing for environmental reasons.[83] Similar proposals covering the OCS were also opposed by industry representatives. During the 1984 debate over whether to remove sections of the OCS to leasing, for example, API and IPAA representatives both argued the 1978 OCS Lands Act should not be changed because it helped "reduce our dangerous dependence on foreign oil."[84]

In contrast, there was no industry agreement on the merits of import barriers. The lack of consensus was noted by IPAA President Lloyd N. Unsell: "I must say that the industry has not come forth with any answers. There has not been a developing consensus, just as there is no political consensus."[85] A series of hearings held in the Senate in 1986 and 1987, when prices fell to their lowest point, provide an opportunity to examine the positions firms and policymakers took towards the question of oil imports. These include hearings held to debate the merits of the Energy Security Act (S 694), introduced by Texan Senator Lloyd M. Bentsen to the Senate in 1987. Although the bill was referred to the Senate Finance Committee where it remained, hearings were held that offered firms, as well as members of Congress and the executive, an opportunity to present their views. The act did not call directly for the introduction of a tariff on imports, but it did require the president to establish a ceiling on oil imports, and prohibited the ceiling

Energy and Agricultural Taxation of the Committee on Finance, US Senate, 100th Congress, 1st session, 1987, 204–6.

83 Statement of Douglass L. Francisco, Senior Government Relations Representative, Independent Petroleum Association of America, on September 23, 1982, to the Hearing Before the Subcommittee on Public Lands and Reserved Water of the Committee on Energy and Natural Resources, US Senate, 97th Congress, 2nd Session, 1982, 207.

84 Statement of Thomas Wright, Western Oil and Gas Association and American Petroleum Institute, on September 20, 1983, to the US House of Representatives Hearings Before the Subcommittee on Mining, Forest Management, and Bonneville Power Administration of the Committee on Interior and Insular Affairs, "Consideration of Legislation to Restrict the Department of the Interior's Oil and Gas Leasing Program on the Outer Continental Shelf," US House of Representatives, 98th Congress, 1st session, 1983, 85.

85 Statement of Lloyd N. Unsell, president of Independent Petroleum Association of America, on September 18, 1986, to the Hearing Before the Committee on Energy and Natural Resources, US Senate, 99th Congress, 2nd session, 1986, 102.

to exceed fifty percent of total crude and oil product consumption in any given year. The act also gave the president the explicit authority to implement a tariff on oil imports, and it was interpreted in committee as seeking the reintroduction of import barriers.

During these hearings, some independents argued in favor of introducing an import fee on crude oil in order to stabilize domestic prices.[86] Smaller associations of independent producers also came out in favor of an import tax, as did the Independent Petroleum Association of New Mexico, which had membership from 600 independent oil and gas producers.[87] The position of the API on import tariffs, on the other hand, was complicated by the greater diversity of its membership. The API president noted these different views stating that an import tax should not be ruled out, but should be subject to careful study given the costs it was likely to impose.[88] In earlier testimony the API was less equivocal, arguing that an import tariff would have severe effects on industries using oil: "In summary, new taxes on energy, depending on their form, will either jeopardize the energy or the economic well-being of this country, or will do both."[89] Surprisingly, the IPAA noted its opposition to the introduction of an oil tariff following a review of the question by a task force created to examine the merits of the question. While their testimony suggests they recognized there may be some economic benefit for domestic producers through the imposition of an import tariff,

[86] Statement of Raymond H. Hefner, Jr., chairman, Independent Petroleum Association of America, on June 5, 1987, to the Hearing Before the Subcommittee on Energy and Agricultural Taxation of the Committee on Finance, US Senate, 100th Congress, 1st session, 1987, 108–21. This position was supported by other independent producer associations. See, for example, the testimony of Robert L. Bayless on behalf of the Independent Producers Association, on February 9, 1987, to the US Senate Hearings Before the Committee on Energy and Natural Resources, 100th Congress, 1st session, 1987, 340.

[87] Statement of Fred J. Schlicher, vice-president of the Independent Petroleum Association of New Mexico, on February 9, 1987, to the US Senate Hearings Before the Committee on Energy and Natural Resources, US Senate, 100th Congress, 1st session, 1987, 270–2.

[88] Testimony of Charles J. DiBona, president of American Petroleum Institute, on January 30, 1987, to the US Senate Hearings Before the Subcommittee on Finance, US Senate,100th Congress, 1st session 1987, 133–53.

[89] Statement of Charles J. DiBona, president, American Petroleum Institute, on June 9, 1982, to the US Senate Subcommittee on Energy and Agricultural Taxation of the Senate Committee on Finance, "Energy Tax Options," 97th Congress, 2nd session, 1982, 55. Independent refiners agreed with the need for greater intervention. Lower oil prices hurt domestic crude production, but increased demand and therefore refinery utilization rates. Their position,

this was balanced against uncertainty about its design, and the possibility that its introduction may also lead to the reduction or abolition of subsidies that were of long-term benefit to members.[90]

The testimonies of individual firms confirm these divisions. Small, domestically focused firms, as well as the large diversified majors, were united in supporting loosening restrictions on the leasing of federal lands.[91] Firms were divided on trade barriers, however. Chevron argued that the federal government should promote production domestically, but not restrict imports.[92] This position was supported by Sun Co., a

like Japanese refiners focused in that market segment and domestically, was that the United States should maintain sufficient capacity to meet domestic needs on national security grounds: "Our oil supply is only the first link in the energy security chain. The second link is the US refining industry. Oil is basically useless to our economy unless it is refined ... Each link in the chain must be secure to the highest possible degree. Each link must be economically self-sustaining to obtain overall energy security." Statement of the Independent Refiners Coalition on January 30, 1987, to the US Senate Hearings Before the Subcommittee on Finance, US Senate, 100th Congress, 1st session 1987, 323. The IRC was made up of 13 firms and associations that focused on the refining segment of the oil production chain. On trade barriers, refiners argued that a gasoline import tax should be introduced if crude imports were taxed, in order to ensure refined products remained competitive within the domestic market. While the proposed instrument differed, they supported implementing barriers to entry on refined products; unsurprising given the common interests of independent refiners in both countries in increasing barriers to entry for imported refined products. Statement of the Independent Refiners Coalition, on January 30, 1987, to the Hearings Before the Subcommittee on Finance, US Senate, 100th Congress, 1st session 1987, 327.

90 "On the Issue of Taxing Petroleum Imports: Statement of the Independent Petroleum Association of America," on February 27, 1986, to the Hearings Before the Subcommittee on Energy and Agricultural Taxation of the Committee on Finance, US Senate, 99th Congress, 2nd session, 1986, 291.

91 Statement of Max Pitcher, vice-president for Conoco, Conoco Statement on S. 2801, on September 23, 1982, to the Hearing Before the Subcommittee on Public Lands and Reserved Water of the Committee on Energy and Natural Resources, US Senate, 97th Congress, 2nd Session, 1982, 199–206; Statement of Richard A. Wolgast, Land Manager, Rocky Mountain Region, Phillips Petroleum Company, on September 23, 1982, to the Hearing Before the Subcommittee on Public Lands and Reserved Water of the Committee on Energy and Natural Resources, US Senate, 97th Congress, 2nd session, 1982, 172; Statement of Gerald F. Rome, vice-president, Texaco USA, on August 29–30, 1984, to the Hearings of the Subcommittee on Fossil and Synthetic Fuels of the Committee on Energy and Commerce, US House of Representatives, 98th Congress, 2nd session, 1984, 221–9.

92 Testimony of J.W. Dewes, General Manager, Chevron Corporation, on March 14, 1986, to the Hearings on Petroleum Industry Issues, "Domestic and

mid-range firm with an international presence. Its chairman, Theodore A. Burtis, gave at times colorful testimony on the national security implications of rising imports in which he argued for the repeal of the Windfall Profits Tax and other tax changes, to improve the incentives for domestic exploration and production, and improved leasing rules for federal lands, but opposed the imposition of an import tariff or quota.[93] Callon Petroleum, on the other hand, which was a small, independent, and domestically focused firm, argued that "the independents and smaller oil companies are paying the price for a healthier domestic economy, and frankly many of them feel taken by the Congress and the Administration." The president of the firm lobbied in favor of restoring the mandatory oil import program established in 1959.[94] This position was supported by Transco Energy, which was a major pipeline owner and operator that also produced oil domestically.[95]

Stable but divided policymaker strategies

These divisions were mirrored in Congress. Legislation was introduced by members from oil-producing districts that sought to reduce crude imports. In the House of Representatives, Bill Richardson of New Mexico's third congressional district proposed an eight dollar a barrel fee on oil and oil product imports in a bill introduced in March 1986 (HR 4454); John Bryant of Texas' fifth district similarly proposed a national import ceiling for oil and oil product imports in legislation introduced on July 29, 1986 (HR 5272). The National Petroleum Security Act (HR 2662) was introduced to the House in June 1987 by

International Petroleum Situation and the Implications of Fees on Imported Oil," Senate Committee on Energy and Natural Resources, US Senate, 99th Congress, 2nd session, 1986, 183.

93 Statement by Theodore A. Burtis, chairman of Sun Company, Inc., on March 20, 1986, to the Hearings on Petroleum Industry Issues, Committee on Energy and Natural Resources, US Senate, 99th Congress, 2nd session, 1986, 269–70.

94 Statement of Alan M. Warren, president of Callon Petroleum Co., Energy Security Policy, on August 29–30, 1984, to the Hearings of the Subcommittee on Fossil and Synthetic Fuels of the Committee on Energy and Commerce, US House of Representatives, 98th Congress, 2nd session, 1984, 57–61.

95 The chairman of the firm, George Slocum, noted his "wholehearted support" for an import fee in the 1986 hearings. Statement of George Slocum, president and Chief Operations Officer, Transco Energy Co., on February 27, 1986, to the Hearings Before the Subcommittee on Energy and Agricultural Taxation of the Committee on Finance, US Senate, 99th Congress, 2nd session, 1986, 270.

Eldon Boulter of Texas' thirteenth district. The majority of the thirty-two co-sponsors of the bill were from oil-producing districts such as Texas, Oklahoma, Colorado, Ohio, and California. Matching legislation was introduced by Senator Lloyd M. Bentsen of Texas in the form of the Energy Security Act (S 694) in March of the same year. The majority of the twenty-five co-sponsors of the bill were once again from oil-producing districts.

Debate within committee confirms the support of these members for an import fee. Senator J. Bennett Johnston from Louisiana advocated for the imposition of an oil import fee: "I know that the imposition of an oil import free would be painful and would wipe out some of the macroeconomic benefits of the oil price decline. But those who eschew an oil import fee must understand the magnitude of our plight."[96] Support for an oil import tax was echoed by Don Nickles, the senator from Oklahoma.[97] In hearings held to discuss the merits of an oil import fee on energy security, Senator Lloyd Bentsen of Texas argued:

> If we fail to respond to the flood of cheap OPEC oil, by putting an oil import fee in place, there's no question that we will soon see the same serious national security implications develop with regard to oil production that we have already seen for refining.[98]

Members from non-oil-producing districts, on the other hand, opposed oil import fees. John H. Chafee of Rhode Island argued that an import tax would impose unfair costs on industry in his state, and on New England as a whole: "In Rhode Island, where our industries already pay more for energy than any other state, the effects of the fee could be devastating – endangering thousands of jobs and pricing our products

[96] Statement of Honorable J. Bennett Johnston, US senator from the State of Louisiana, on March 20, 1986, to the Hearings on Petroleum Industry Issues, Committee on Energy and Natural Resources, US Senate, 99th Congress, 2nd session, 1986, 193 and 264.

[97] Statement of Honorable Don Nickles, US senator from the State of Oklahoma, on March 20, 1986, to the Hearings on Petroleum Industry Issues, Committee on Energy and Natural Resources, US Senate, 99th Congress, 2nd session, 1986, 193 and 258.

[98] Statement by Senator Lloyd Bentsen, Senate Finance Committee Hearings, February 27, 1986, 14.

out of world markets."[99] Maine Senator George J. Mitchell also argued trade barriers were being debated because "the oil producing states want to find a way to bail out the oil industry from the steep price decline of recent weeks."[100] This echoed the position of representatives from the chemicals, steel, trucking, and agriculture industries, as well as marketers of gasoline, who uniformly opposed the imposition of a tariff in hearings.[101]

The executive branch also stated that it would not countenance a tariff on imported crude and petroleum products.[102] Multiple rationales were offered by administration officials for the position. The general commitment of the administration to economic liberalization appears to have been one motivation. Secretary of Energy James B. Edwards made an explicit link between the preferred policy strategy of the Reagan administration and economic liberalization more broadly:

> [Regulation] has only been an experiment – and it has proved not to be a very good one, especially in the field of energy. The free market allows consumers and the entrepreneur to experiment and to make mistakes. But it doesn't subject the whole country to the alchemy of planners in Washington.[103]

There is also evidence that the White House recognized that the problem of energy security had changed because of changes in market

[99] Statement of Senator John H. Chafee on February 27, 1986, to the Hearings Before the Subcommittee on Energy and Agricultural Taxation of the Committee on Finance, US Senate, 99th Congress, 2nd session, 1986, 8.

[100] Statement of Senator George J. Mitchell on February 27, 1986, to the Hearings Before the Subcommittee on Energy and Agricultural Taxation of the Committee on Finance, US Senate, 99th Congress, 2nd session, 1986, 22.

[101] See, for example, "Impact of Imported Petroleum Products on the Domestic Petroleum Industry," on June 4, 1985, to the Hearing Before the Subcommittee on Energy Regulation and Conservation of the Committee on Energy and Natural Resources, US Senate, 99th Congress, 1st session, 1985.

[102] Patrick Crow, "Herrington: No Government Bailout for US Oil Industry," *Oil & Gas Journal*, May 5, 1986: 61.

[103] Ibid. The view that equated liberalization with energy security was also offered by the Assistant to the President for Communications David R. Gergen in a briefing made to reporters on March 22, 1982: "the forces of a free marketplace are now having a positive impact upon America's energy outlook. The substantial increases we are now seeing in the search for domestic oil and in the achievement of greater energy efficiency both

structure. Secretary of Energy William Martin argued, for example, that rising imports were problematic both for domestic industry and for energy security, but that imports were not a useful indicator of US energy security.[104] This view was echoed in Reagan's response to Congress' determination that rising oil imports represented a threat to US national security: "Given the nature of the international oil market, the United States alone cannot assure its energy security. Consequently, we continue to work closely with our partners in the International Energy Agency (IEA) to improve our mutual energy security."[105] Finally, the administration also made reference to the trade-offs between protecting domestic industry and the costs this would impose on industry and consumers.[106]

Together, this represented a powerful coalition arrayed in opposition to the imposition of an import tariff. Testimony suggests that opposition in Congress and among industry to the continuation of subsidies, on the other hand, was weaker, and oil firms were unified in favor. It is unsurprising, therefore, that while the Tax Reform Act of 1986 made changes to the incentives provided to the domestic oil and gas industry, it retained the tax deductions for drilling costs and depletion allowance that industry supported.

Case summary

Oil market volatility in the 1980s harmed domestic industry, as it did in Japan and France. The position of firms, on the other hand, was divided. Small and poorly diversified independents focused on producing in the United States were particularly hurt. Many of these firms, and the industry association that represented them, responded by seeking the support of the government, as did Japanese firms. While the IPAA opposed import controls, however, a number of domestically

illustrate the power of market incentives." Woolley and Peters, *The American Presidency Project.*

104 Statement of William F. Martin, Deputy Secretary, Department of Energy, on Energy Security Act of 1987, on June 2, 1987, to the Hearing Before the Subcommittee on Energy Regulation and Conservation of the Committee on Energy and Natural Resources, US Senate, 100th Congress, 1st session, 1987, 27.

105 Woolley and Peters, *The American Presidency Project.*

106 Statement of William F. Martin, Deputy Secretary, Department of Energy, on Energy Security Act of 1987, on June 2, 1987, 78.

focused firms supported the imposition of a tax on oil imports. Large, integrated firms, on the other hand, opposed this measure, just as French firms were disinterested in maintaining barriers to imports. These differences in the strategies adopted by firms can readily be explained by differences in firm characteristics. US firms, on the other hand, were united in seeking the abolition of the Windfall Profits Tax, and increasing access to federal lands and the OCS. This differs from both the French and Japanese cases, which is unsurprising given their lack of domestic resources.

To what extent were these firm strategies represented in outcomes? Congressional opinion was similarly divided. Representatives in a number of oil-producing regions sought a fee on oil imports and introduced legislation requiring the president to act if oil imports increased. Representatives from states with a heavier manufacturing base, on the other hand, argued against oil import barriers. The White House also refused to countenance an oil import tax, but agreed to more limited measures. Outcomes reflected these divisions. Policymakers repealed the Windfall Profits Tax and maintained subsidies designed to promote industry profitability in the tax package signed into law in 1986. Industry and supportive congressional members failed, on the other hand, to persuade the White House or other members to impose barriers on oil imports.

Case study 2 – rising prices and the George W. Bush administration

Oil price volatility in the 2000s renewed debate over how best to govern the US oil market. In the presidential campaign, candidate George W. Bush proposed a series of supply-side measures in response to rising prices. This matched the Republican Party position in favor of increasing domestic production of oil, gas and coal.[107] It also matched the interests of firms, which supported increasing the incentives for domestic production as they did in the 1980s. The most contentious component of the proposal was the inclusion of drilling in the Arctic

[107] Staff reporter, "Energy a Hot-Button Issue; in US Presidential Election Campaign," *Oil & Gas Journal*, October 9, 2000: 20; Staff reporter, "GOP Platform Calls for Strong US Energy Policy," *Oil & Gas Journal*, August 14, 2000: 30.

National Wildlife Reserve (ANWR). Democratic Party nominee Gore's position focused on conservation measures and opposed oil development in the ANWR, but included limited measures to increase domestic production of oil and gas.[108]

Once the president took office he ordered Vice-President Cheney to head a Cabinet Task Force, including the Secretaries of Treasury, Energy, Commerce, Interior, Transport, and Agriculture, to draft a national energy strategy.[109] Released on May 17, 2001, the National Energy Plan (NEP) proposed subsidies for enhanced oil recovery technologies, increasing deep-water and marginal production royalties, and expediting exploration and production in the OCS and coastal zone.[110]

The emphasis on domestic production was echoed in legislation proposed in Congress. Alaska Republican and Chairman of the Senate Energy and Natural Resources Committee Frank Murkowski joined with the ranking Democrat, Louisiana representative John Breaux, to introduce legislation enabling exploration and production in the ANWR, and increasing incentives for exploration and drilling on federal lands. The bill also proposed halting royalty payments during periods of low oil prices and establishing a tax credit of three dollars per barrel for production from marginal wells. The content of the Murkowski–Breaux Bill closely mirrored that of the White House plan; the Energy Secretary stated that thirty of the recommendations made in the administration's NEP were reflected in the Murkowski Bill.[111] House legislation proposed similar measures. Billy Tauzin of Louisiana, Chairman of the House Committee on Energy and Commerce, introduced the Securing America's Future Energy Act of 2002 (HR 4), along with representatives from Utah, Oklahoma, and California.[112] The legislation

108 Staff reporter, "Bush, Gore Highlight Energy Strategies in Debate," *Oil & Gas Journal*, October 9, 2000: 24.

109 Staff reporter, "US Oil and Gas Industry Has High Expectations from Bush Administration on Energy Issues," *Oil & Gas Journal*, February 12, 2001: 66.

110 Staff reporter "Energy Policy to Dominate US Political Agenda," *Oil & Gas Journal*, May 28, 2001: 20.

111 Statement of Honorable Spencer Abraham, Secretary, Department of Energy, on May 24, 2001, to the Hearings Before the Committee on Energy and Natural Resources, US Senate, 107th Congress, 1st session, 2001, 10. For a summary of the measures see Staff reporter, "Industry Praises Murkowski's Omnibus Energy Bill," *Oil & Gas Journal*, March 5, 2001: 40.

112 The legislation was prepared in coordination with the White House. See "National Energy Policy Report of the National Energy Policy Development

included measures to promote production from marginal wells and tax credits for enhanced oil recovery, in addition to support for conservation and renewables. It was introduced to the House on July 27, 2001, and passed on August 2 on a vote of 240 to 189. The Democratic Party supported more limited measures to increase domestic production. Congressional Democrats also announced their opposition to drilling in the ANWR, but supported tax benefits for independent producers of oil and gas.[113]

The progress of energy legislation through Congress was slowed by the terrorist attacks of September 11, 2001. Both the House and the Senate took up the debate over energy once again in 2003. In June 2004 the House passed HR 4503 by a margin of 244 to 178, which largely reflected the proposals of the Bush administration.[114] In the Senate the bill stalled over the treatment of an issue unrelated to the problem of domestic production – liabilities for water contamination from the gasoline additive methyl tertiary butyl.[115] Legislation finally passed both houses in 2005. The House version of the bill (HR 6) included 8.1 billion dollars in tax incentives aimed at boosting domestic energy production, and passed on a vote of 249 to 183. The Senate passed the bill by a vote of 85–12, and also included measures to promote domestic oil production.

The bills were reconciled in conference in August 2005. Provisions to boost domestic production remained in place. The conference between the two houses finished work on July 26, with the House voting 275–156, and the Senate voting 74–26. The signing of the legislation by President Bush in August 2005 marked the end of a four-year process, and was complemented by a 2006 law expanding the area of the OCS available for drilling. The final legislation included a wide range of measures, including incentives to improve energy efficiency, renewable energy subsidies, and programs to support the coal industry. It also included production incentives for oil as demanded by industry.

Group," on June 13, 2001, to the Hearing Before the Subcommittee on Energy and Air Quality of the Committee on Energy and Commerce, House of Representatives, 107th Congress, 1st session, 2001, 7.

113 Maureen Lorenzetti, "House Democrats Unveil Their Own Energy Plan," *Oil & Gas Journal*, May 21, 2001: 30.

114 Maureen Lorenzetti, "US House Repasses Sweeping Energy Legislation," *Oil & Gas Journal*, June 21, 2004: 28.

115 Maureen Lorenzetti, "Senate GOP Leaders Affirm Energy Bill Dead for the Year," *Oil & Gas Journal*, December 1, 2003: 27.

Stable and unified firm strategies

Industry united in favor of easing access to federal lands and the OCS, as well as increasing subsidies, consistent with their position in the 1980s. Unlike the 1980s, however, when falling prices led some firms to shift in favor of import controls, congressional testimony suggests there was little demand from firms for imposing limits on imports under conditions of rising prices. One possible reason for this is that firms performed better under conditions of high prices. This was the case both for the large, diversified majors and for independents. CEO and Chairman of Exxon Mobil, Lee Raymond, for example, noted that the firm had achieved its second highest earnings in company history in nominal terms in the financial year 2001, with a return on capital of eighteen percent.[116] Earnings in the financial year 2004 stood at 25.3 billion dollars, which was the highest ever recorded by the firm in nominal terms. The firm achieved a return on capital of twenty-four percent, and a return in the production and marketing of oil and gas of thirty-three percent from operations in thirty-seven countries.[117] Chevron had a return on capital employed of 17.3 percent in 2000, and 14.7 percent in 2001 after merging with Texaco once the merger-related costs were excluded.[118] In the financial year 2005, Chevron returned 21.9 percent on capital employed.[119] While financial data on smaller, independent firms is less available, the number of oil rigs employed offers one way of assessing the health of the industry: after falling to a low of 625 rigs in 1999, the number of rigs used increased to more than 1,000 in 2003, rising to 1,381 in 2005, suggesting firms were increasingly bullish about their prospects.

This did not make government irrelevant. In testimony, the API – which had the broadest membership of firms across all segment of the industry – supported government intervention, arguing that promoting domestic supplies of crude oil was "an enormous challenge [and] not totally within the industry's control. Government also has an

[116] Exxon Mobil Corporation, *Exxon Mobil Annual Report 2001* (New York: Exxon Mobil Corporation, 2001), 3.

[117] Exxon Mobil Corporation, *Exxon Mobil Annual Report 2004* (New York: Exxon Mobil Corporation, 2004), 2 and 9.

[118] ChevronTexaco Corporation, *ChevronTexaco 2001 Annual Report* (San Ramon, CA: ChevronTexaco, 2001), 9.

[119] Chevron Corporation, *Chevron Annual Report 2005* (San Ramon, CA: ChevronTexaco, 2005), 4.

important role to play."[120] In written comments to the hearings to the 2001 legislation introduced by Senator Murkowski, the API argued for removing regulatory barriers to increasing production within the United States.[121] It focused in particular on immediate steps that could be taken to increase domestic production, presenting proposals for speeding up permitting on lands already available for drilling, and relaxing regulatory constraints on OCS.[122] The IPAA agreed with the API position, supporting access to federal lands through Sale 181, which promised to enable leasing on an area of the Gulf of Mexico estimated to contain up to 1.9 billion barrels of oil, as well as measures to speed permitting of onshore leasing.[123]

Tax benefits were also supported by industry. In Senate hearings the IPAA argued for renewing the Deepwater Royalty Relief Act, and for new incentives for operations carried out in wells below 15,000 feet.[124] Jerry Jordan, president of Jordan Energy and IPAA chairman testified on behalf of the IPAA, the National Stripper Well Association, and thirty-two cooperating state and regional associations, calling for access to capital through improving incentives to drilling in the tax code, and access to federal non-park lands.[125] Charles N. MacFarlane,

[120] Statement of Red Cavaney, president and CEO, American Petroleum Institute, Oil Supply and Prices, on February 13, 2003, to the Hearing before the Committee on Energy and Natural Resources, US Senate, 108th Congress, 1st session, 2003, 18.

[121] Statement Of American Petroleum Institute, on May 24, 2001, to the Hearings Before the Committee on Energy and Natural Resources, US Senate, 107th Congress, 1st session, 2001, 181–7.

[122] Statement of Mark Rubin, Upstream General Manager, American Petroleum Institute, on May 22, 2001, to the Oversight Hearing for the Subcommittee on Energy and Mineral Resources of the Committee on Resources, US House of Representatives, 107th Congress, 1st session, 2001, 10–17.

[123] Statement of Tom Young, vice-president of Business Development, Mariner Energy Inc., on May 24, 2001, to the Hearings Before the Committee on Energy and Natural Resources, US Senate, 107th Congress, 1st session, 2001, 138–41. See also Staff reporter, "Industry Praises Murkowski's Omnibus Energy Bill," *Oil & Gas Journal*, March 5, 2001: 40.

[124] Statement of Tom Young, vice-president of Business Development, Mariner Energy Inc., on May 24, 2001, to the Hearings Before the Committee on Energy and Natural Resources, US Senate, 107th Congress, 1st session, 2001, 143–5.

[125] Statement of Jerry Jordan, president of Jordan Energy, National Energy Policy, on February 28, 2001, to the Hearing Before the Committee on Energy and Commerce, US House of Representatives, 107th Congress, 1st session, 2001, 70–6.

a tax specialist from Chevron testifying on behalf of the API and US Oil and Gas Association, argued that "the tax treatment of the financing and structuring of these ventures is one of the essential elements of decisions whether to proceed."[126]

Testimony from individual firms mirrored the positions adopted at the two major industry associations. This was the case both for larger, diversified firms, as well as the smaller independents. Jim L. Bowles of Phillips Petroleum Company argued in favor of improving access to federal lands: "we are willing to make enormous investments to meet these ends, but we must have access to our natural resources for exploration and production."[127] In hearings on the OCS, the Exploration Manager from Shell argued that long lead times in the industry needed certainty from the federal government on leasing schedules.[128] The Chevron representative argued that developing the OCS would reduce the need for foreign imports of oil, and also argued that government incentives were crucial in stimulating firm activities in deep-water, given the high risks involved in the venture.[129]

Stable but divided policymaker strategies

Testimony suggests that the measures promoting oil production included in the final bill signed by the president were consistent with

[126] Statement of Charles N. MacFarlane, Assistant General Tax Counsel, Chevron Corporation, on behalf of the American Petroleum Institute, Domestic Petroleum Council, and US Oil and Gas Association, on June 31, 2001, to the Hearing Before the Subcommittee on Select Revenue Measures of the Committee on Ways and Means, US House of Representatives, 107th Congress, 1st session, 2001, 43.

[127] Statement of Jim L. Bowles, president, Americas Division, Phillips Petroleum Company, March 7, 2001, to the Oversight Hearings Before the Committee on Resources, US House of Representatives, 107th Congress, 1st session, 89–90.

[128] Statement of Greg Guidry, Exploration Manager, Shell Exploration and Production Company, on August 13, 2001, to the Oversight Field Hearing Before the Subcommittee on Energy and Mineral Resources of the Committee on Resources, US House of Representatives, 107th Congress, 1st session, 2005, 49–50.

[129] Statement of Paul K. Siegele, vice-president, Deepwater Exploration and Projects, Chevron North America, on January 25, 2007, to the Hearing Before the Committee on Energy and Natural Resources, US Senate, 110th Congress, 1st session, 2007, 35.

the demands of firms. Many of the supporters of the legislation also received substantial support from industry. In the electoral cycles from 2000 to 2006, for example, oil and gas-related political action committees gave a mean of more than 6.5 million dollars to federal candidates in nominal terms. On the other hand, congressional members from oil-producing states also had an incentive to supply these policies that was independent of the demands from firms. The largest oil-producing state, Texas, had more than 150,000 producing oil wells and 250,000 employees (three percent) in the sector.[130] Of the sponsors of the initial senate legislation, Louisiana had more than 17,000 producing wells, with more than 90,000 employees (five percent of total non-farm employees) of the workforce directly employed in the production, refining, transport, or sale of petroleum in 2004, as well as almost 400 million dollars in tax revenue from oil production.[131] Alaska had almost 2,400 producing wells and 12,300 employees (2.5 percent of total population in 2000) working in the sector, but with the likelihood of substantial increases in employment, as well as state revenues, if oil production in the ANWR became possible.[132] It is unsurprising, therefore, that senators from oil-producing states supported greater production. Senator Craig Thomas from Wyoming argued that allowing leasing within Area 181 in the Gulf of Mexico was a useful response to lowering US dependence on the international market for the supply of oil, in addition to increasing fuel efficiency and other demand-side measures.[133] Senator Murkowski from Alaska also promoted the benefits of promoting oil and gas production from the OCS, as well as within Alaska in the ANWR and the Arctic

[130] Calculated from IPAA, *IPAA Oil and Gas Producing Industry in Your State 2005–2006* (Washington, DC: IPAA, 2007), 75, and data from the Labor Market and Career Information Department of the Texas Workforce Commission.

[131] Calculated from Louisiana Department of Labor, *Louisiana Employment and Wages Annual Report 2004* (Baton Rouge, LA: Department of Labor, 2005), 2; IPAA, *IPAA Oil and Gas Producing Industry in Your State 2005–2006*, 41. Tax data calculated using Louisiana Department of Natural Resources data, and include both state revenues from production and OCS.

[132] Employment data from IPAA, *IPAA Oil and Gas Producing Industry in Your State 2005–2006*, 21.

[133] Statement of Honorable Craig Thomas, US senator from Wyoming, on March 7, 2006, to the Hearing Before the Committee on Energy and Natural Resources, US Senate, 109th Congress, 2nd session, 2006, 12–13.

coastal plain.[134] Senator Orrin Hatch of Utah strongly advocated for the benefits of increasing the production of unconventional oil from shale and tar sands in order to "promote a vision of the future that is less dependent on oil."[135]

The position of congressional members was once again not unified, however. As congressional supporters sought to move legislation through Congress they faced opposition over allowing oil and gas exploration and production in the ANWR and other issues. A number of the important committees for energy-related legislation were relatively unified. The House Energy and Commerce Committee, for example, agreed to legislation put to the House, which was more supportive of oil interests than the Senate version of the bill, by a vote of fifty for versus five members against (HR 2587). The outcome in committee was less lopsided in 2003 and 2005 when legislation came up for review once again, but showed a strong majority in favor of the legislation; in 2003 the committee passed HR 1644 by a margin of 36–17, and in 2005 it passed HR 6 by a vote of 39–16.[136] The House Resources Committee also considered parts of the legislation focusing on the management of federal lands. Debate was heated because of the issue of the ANWR. The committee nevertheless voted in support of drilling in 2001 by a margin of 26–17, and again in 2003 by a margin of 32–14. Finally, on the House side, the House Ways and Means Committee considered the components of the legislation related to taxation, approving increased subsidies by a vote of 24–17 (HR 2511). In 2003 it passed by a voice vote a version of the legislation that included 18.7 billion dollars in tax benefits for the energy industry, including oil, and voted in favor of legislation resembling the final legislation by a vote of 26–11 in 2005.

Not all congressional opinion promoted the merits of increasing domestic oil production. Committee negotiations saw the demands of these members whittled away by those who disagreed with promoting the production of oil domestically. Senator Coleman from Minnesota, for example, emphasized the national security risks associated with

[134] Statement of Honorable Linda Murkowski, US senator from Alaska, on April 12, 2006, to the Hearing Before the Committee on Energy and Natural Resources, US Senate, 109th Congress, 2nd session, 2006, 2–3.

[135] Statement of Honorable Orrin Hatch, US senator from Utah, on March 7, 2006, to the Hearing Before the Committee on Energy and Natural Resources, US Senate, 109th Congress, 2nd session, 2006, 51.

[136] All data drawn from the CQ Almanac.

relying on external markets for the supply of oil, but proposed that the best response was to increase the domestic production of biofuels. Other senators focused on demand-side responses.[137]

In the Senate, the Energy and Natural Resources Committee played an important role in considering versions of the legislation. In 2001 it moved towards completing work on the bill, but committee work was halted by the Democratic leadership, which opposed drilling in the ANWR. In 2003 the committee was evenly split, approving the Senate version of the bill (S 1005) by a margin of 13–10, while in 2005 the vote was more decisive, favoring the proposed legislation by a margin of twenty-one for and just one against. Examined through these committee debates, the views of congressional members suggest that the major division was between pro-production members of Congress from areas with higher levels of oil and gas production, and those that had fewer hydrocarbons within the districts and states they represented.[138]

Case summary

The Energy Policy Act was signed into law by President Bush on August 8, 2006. The legislation took four years to pass through Congress, and did not include all the proposals demanded by industry. It nevertheless included measures to increase domestic oil production, from tax incentives and subsidies totaling 14.6 billion dollars, with 2.8 billion dollars set aside for the production of fossil fuels, and authority conferred to the Interior Department to conduct a survey of prospective resources in the OCS that could be made available for domestic drilling. The president noted his satisfaction with the outcome in remarks made after signing the bill into law: "I'm confident that one day Americans

[137] Statements of the Honorable Evan Bayh, US senator from Indiana, Norm Coleman, US senator from Minnesota, on March 7, 2006, to the Hearing Before the Committee on Energy and Natural Resources, US Senate, 109th Congress, 2nd session, 2006, 3–7.

[138] Statement of Honorable Pete V. Domenici, US senator from New Mexico, on February 13, 2003, to the Committee on Energy and Natural Resources, US Senate, 108th Congress, 1st session, 2003, 2. Montana Senator Conrad Burns followed Domenici in calling for increasing production from within the US, emphasizing the exploration and production opportunities that existed in federal lands. See Statement of Honorable Conrad Burns, US senator from Montana, on February 13, 2003, to the Committee on Energy and Natural Resources, US Senate, 108th Congress, 1st session, 2003, 4–5.

will look back on this bill as a vital step toward a more secure and more prosperous nation that is less dependent on foreign sources of energy."[139]

Industry was more united in the debate than in the 1980s. Unlike the 1980s, when falling prices led some independent firms to support the imposition of import barriers in addition to tax breaks, high oil prices meant there were few demands for government intervention in the form of trade barriers. A complicated set of trade-offs involving the environmental as well as economic effects of intervention were also a source of disagreement between members of Congress, as legislation made its way through the committee phase and onto the floor. Supporters of production were nevertheless successful in ensuring some of the measures they supported emerged from this process and were incorporated in the final legislation that reached the desk of the president.

Summary of changes in oil market governance in the United States

There continues to be limited but meaningful intervention by the US government in order to increase the share of oil produced within national borders as a ratio of total demand. The scale of this intervention does not approach that of the 1950s–1970s, when quotas were used to increase the price of internationally produced oil relative to oil produced at home. Yet energy independence remains a component of presidential and congressional policy strategies.

This position has been consistently supported by domestic industry. Independent firms with a majority of domestic production have lobbied to increase the incentives for exploration and production of crude oil within the continental United States. As prices fell in the 1980s some supported reimposing an oil import tariff in order to improve competitiveness. They were joined by congressional members from oil-producing districts. The ability of these firms and their congressional supporters to achieve their preferred outcome was limited, however, by the lack of support from the large integrated majors. It was also

[139] Jim VandeHei and Justin Blum, "Bush Signs Energy Bill, Cheers Steps Toward Self-Sufficiency; Measure Includes Billions in Tax Breaks for Industry," *Washington Post*, August 9, 2005: A03.

limited by the lack of support from congressional representatives concerned about the costs import tariffs would impose on consumers of oil products.

How well does the framework developed in Chapter 3 explain outcomes? The existence of substantial reserves of oil within the domestic market makes the politics of oil in the United States different from the other cases considered in this study. Most importantly, the existence of abundant domestic resources of crude oil affected firm characteristics independently from the effect of the industrial compacts negotiated in earlier periods. As the first part of this chapter records, the domination of the international market by US production in the years before World War II not only led to the emergence of export-oriented and competitive firms, although that is certainly one outcome. It also led to a bifurcated industry structure, with integrated and internationalized majors operating alongside thousands of smaller independents focused on oil production alone. Evidence nevertheless suggests the industrial compacts negotiated between firms and government in the case of the United States had an effect. Most importantly, the consistent support of Congress through tax credits and subsidies, as well as import barriers, contributed to the ongoing importance of independent producers to the political economy of oil in the United States.

Our expectations about what firms are likely to demand from policymakers are largely confirmed in the US case, although the existence of domestic resources within the United States made the government useful to larger, more diversified firms. These firms opposed import barriers in response to falling prices, just as French firms were disinterested in maintaining restrictions on imports. A number of independents, on the other hand, shifted to support the imposition of trade barriers as profitability declined in the face of falling prices, mirroring the position taken by Japanese refiners who supported retaining barriers to imports of oil products. One puzzle here is the position adopted by the IPAA, which represents independent producers and opposed the imposition of an import tariff. Testimony in Congress suggests that the rationale for this opposition was not because it saw no benefit in such a measure in isolation, but rather because it expected other measures to be implemented in tandem that could, on balance, hurt members' interests.

The divisions among firms were replicated within Congress. Although a voting record is not available allowing us to assess the position of congressional members because oil-related votes were commonly bundled with other tax issues, or with other energy polices, congressional testimony suggests members from districts with oil production supported the position of firms towards increasing subsidies for firms. Indeed, testimony from the 1980s from a number of members suggests they were willing to go further than the IPAA in pushing for the imposition of a tariff on oil imports. This support was balanced, however, against the concerns of members from the northeast and other areas that lacked domestic oil interests and feared the economic effect of increased prices on industry and consumers in their districts. The collective decision in both the cases examined above settled on the provision of subsidy support for firms, but no imposition of costs through a tariff on oil or oil product imports. In this sense the outcomes in the Japanese and US cases are similar, with the retention of subsidy support for firms but the rejection of import barriers, although Japanese firms operating within Japan were focused on refining and distribution, while smaller and less diversified US firms were focused on oil production within national borders.

One important difference between the French and Japanese, and the US cases, not accounted for in Chapter 3, stems from the existence of domestic oil resources in the United States. This made oil production within the US attractive not only to the smaller, less diversified firms, but also to the majors seeking to diversify operations. Indeed, as we saw in the French case, the United States was an important location for production in the wake of the oil market nationalizations of the 1970s. Thus, while US firms were at times divided on the question of imports, they were united in support of measures to ease access to federal lands and the OCS.[140] This is not an outcome relevant to the other two cases, given their lack of domestic resources.

Alternative explanation – ideational change

There is some evidence to support the contention that learning played a role in driving outcomes. This is most notable in the case of the Reagan administration. Representatives from the administration repeatedly

[140] The importance of intra-industry divisions in weakening the bargaining power of firms is also noted by Milner across a larger number of firms. Helen Milner, "Trading Places: Industries for Free Trade," *World Politics* 40 (1988): 374.

referred to their belief that energy independence by itself was meaningless when judging the risks the United States faced in oil markets. In contrast to the 1950s, they steadfastly refused to heed calls from independents and congressional members from oil-producing districts for the reimposition of trade barriers in order to protect domestic production. Secretary of Energy Abraham, in describing the National Energy Plan of the new administration, argued that a long-term strategy to increase supply was needed, including non-conventional sources such as biomass, solar, and wind power, as well as increasing domestic sources of oil and gas.[141] The administration rejected the idea that promoting domestic production represented a means for promoting self-sufficiency, however. Rather, Abraham argued that "free markets are at the heart of ... a healthy international energy system," and that the development of resources primarily benefitted economic development, rather than enhancing energy security.[142]

Yet if learning mattered in the case of the United States, it did not define the strategies of all policymakers. Evidence shows that members of Congress sympathetic to the interests of the oil sector strongly advocated for greater state intervention as oil prices collapsed in the 1980s, including demanding in committee the imposition of oil import tariffs. They were also able to achieve some of their preferred outcomes. The evidence also suggests not only that policymakers were not unified in their view of the risks associated with relying on the international market for oil, and the appropriate policy response, but that industry also had some influence on outcomes. This suggests in turn that while learning may have played a role in determining outcomes in the United States, it did not dominate outcomes.

[141] Statement of Honorable Spencer Abraham, Secretary, Department of Energy, on May 24, 2001, to the Hearings Before the Committee on Energy and Natural Resources, US Senate, 107th Congress, 1st session, 2001, 8.

[142] Statement of the Honorable Spencer Abraham, Secretary, US Department of Energy, on June 20, 2002, to the Hearing Before the Committee on International Relations, House of Representatives, 107th Congress, 2nd session, 2002, 10–11.

7 Firms, governments, and oil market governance

In the past three decades, governments in the advanced industrialized states transformed their approach towards promoting growth and employment, and managing risk. In this book I showed that the oil sector was not immune from these changes. Governments largely rejected a more interventionist approach to governing domestic oil markets, replacing it with a new strategy that focuses on strategic petroleum reserves as a physical hedge against supply disruptions, but otherwise leaving supply and demand to the market. I also showed the process of liberalization was partial: of the three cases that make up this study, both Japan and the United States retained, and in Japan's case reenergized, support for domestic firms.

These changes in oil market governance are surprising. For almost a century governments have framed oil as qualitatively different from other goods because of the importance of fuels derived from it for industrial growth and military competence. Yet the long-term response of governments to the loss of control of many of the world's most productive oil fields by firms based in the United States and Europe was to reduce, rather than increase, intervention in domestic oil markets, even as governments in some countries continued to try to shape oil markets in order to promote the competitiveness of domestic firms.

In this book I offered a firm-centered explanation to account for this outcome. I focused in particular on how differences in the characteristics of oil firms across countries affected their responses to this transformation of the oil market, including the types of demands they made on policymakers. In contrast to weak and poorly diversified Japanese firms, I found that French firms grew increasingly competitive internationally both by expanding their refining operations outside France into the United States and across Europe, as well as by increasing the number of countries in which they explored for and produced oil. This trend began in the 1960s and was propelled forward by the changes in the structure of the international oil market in the 1970s;

nationalization led the management of both firms to seek new oil fields internationally, and stagnant demand for oil products within France led them to invest in new refineries in Eastern Europe and elsewhere. The French market, and the protection afforded to them by the state, therefore became less important to these firms, and their position towards protection and government ownership shifted as a result.

Japanese firms, on the other hand, remained small and poorly diversified, despite similar initial conditions to France. This led them to continue to focus on the resources available within government as they sought to increase international competitiveness. They did not receive everything they demanded. In the wake of price volatility, refining firms were unable, for example, to defend the import barriers that had protected their competitive position for decades in the face of demands from manufacturers and others for reducing the costs of inputs, and the failure of the existing strategy for supporting firms. Government indebtedness also led to the privatization of some assets. Subsidies, on the other hand, which supported investments made internationally, remained in place, both because firms continued to demand them and because policymakers saw little reason to discontinue this less costly form of support.

A complete explanation for the changes we see in the governance of oil markets also requires us to consider the incentives for policymakers to supply support to firms. In addition to the interests and strategies of oil firms, I showed that while one goal of initial efforts in France and Japan to shape distributional outcomes in oil markets was to regulate the market power of the oil majors, and particularly Standard Oil, a second important characteristic of oil is that products derived from it are important inputs in the transportation sector, and also in a number of manufacturing processes. Political support for firms operating in the oil sector was therefore balanced against the demands from consumers of oil products not to increase the price of inputs.

Industrial compacts as a source of firm characteristics

Approaches to the study of liberalization tend to focus on underlying factor endowments as an important source of firms' and industries' competitive advantages cross-nationally, and therefore the positions they adopt towards economic liberalization. This approach is untenable in the oil sector. Almost all the major advanced industrialized states are poorly endowed with oil resources. Even in the United States, which continues to produce a significant share of world oil, the costs of production are higher than in the Middle East. US producers are therefore reliant to a significant degree on the exercise of market power by OPEC, as the effect of the collapse of oil prices in 1986 clearly shows. Further, differences between firms make a real difference to the types of policies they support, and hence policy outcomes.

I argued that industrial compacts negotiated in earlier periods played an important role in shaping firm characteristics, and therefore the types of demands they made on policymakers. In the case of France, the industrial compacts negotiated between business and government led to the development of firms that responded to the changed balance of opportunities and risks in the international oil market by rejecting the existing structure of oil market governance. Rather than leading to stasis, therefore, increasing returns to policies implemented in the French oil sector led to transformative change, as French oil firms became disinterested in expending resources to retain the industrial compact that had been negotiated in the previous decades and began to lobby against it. This contrasts with the case of Japan, where the compact negotiated between firms and policymakers was an important reason firms remained small and poorly diversified, which led them to continue to demand protection and subsidies from the state. History therefore mattered, shaping the characteristics of firms in ways that conditioned the way they sought to manage changes in the opportunities and risks they faced in the international oil market. Indeed one way to understand the changes made to Japanese institutions governing the oil market is as a renegotiation of the terms under which to promote firm interests, given that the existing institutions had proven incapable of producing a vertically integrated and internationally competitive oil major.

I also sketched and examined the explanatory power of two alternative ways for accounting for the changes we observe in oil markets. The first focuses on the role of ideas in shaping the strategies adopted by policymakers towards oil market regulation. Changes in the laws and policies governing oil markets in Japan, France, and the United States followed a shift in the model used to understand the nature of oil supply in the international market. This in turn was influenced by the transformation in the structure of the international oil market that served to reduce the risks of interruptions of supply.

The nationalizations of oil production by many of the governments in OPEC not only had the effect of transferring ownership of the major centers of crude oil production outside the United States from international oil companies to national governments, it also separated ownership of production of crude oil from ownership over the majority of the refining and distribution infrastructure. Coupled with the rise in production outside the OPEC region this helped create a secondary market for crude oil, replacing a system in which non-US production was largely managed by an oligopoly centered on the major integrated oil firms. These changes also reduced the likelihood of supply disruptions in oil. Given this, it is plausible that policymakers transformed the institutions governing oil markets to better match the changed structure of risk.

In each of the chapters I found evidence that policymakers mattered in renegotiating the industrial compacts used to govern oil markets in the post-1970s environment. Yet rather than ideational change, policymakers were more concerned about the economic and political costs of oil price volatility, and the effect of existing laws and policies on domestic firms following these changes in the international structure of oil supplies. Further, while their views shaped outcomes, they did not determine them. Instead, they were forced to bargain with firms and other actors while seeking to pursue their interests. While a focus on policymakers and how they also responded to the transformation of the international oil market is an important part of explaining outcomes, therefore, it was not ideational change but rather concerns about inflation, and the health of the domestic firms, that most often drove interest in renegotiating existing laws.

A second alternative argument focused on the role of European institutions. European institutions consistently promoted greater liberalization in energy markets. They had been unsuccessful in affecting the policy of member states in the decades preceding the oil shocks, and focused on assessing problems of the viability of the refining sector in the immediate years after oil price volatility in the 1970s led to substantial problems with overcapacity across member states. In the 1980s Europe increased pressure on member states to reduce barriers to trade and investment in energy markets. In the case of France, the Commission made a number of specific recommendations to the French government, including rewriting the national law and reducing barriers to investment.

While the timing of French institutional changes and the recommendations made by European institutions suggest a correlation between these events, the evidence suggests that it was the changing strategy of firms themselves – coupled with changes in policymakers' approach to regulating the economy – that was causal. By the time recommendations were made to the French government to change intervention in the oil market, the system that had previously protected the domestic market and provided capital and market share to national firms was no longer important to firm success. Indeed, firms themselves had pushed for greater liberalization when intervention by the state interfered with their ability to increase profitability.

How confident can we be that these findings travel to other contexts? The case of the United States, examined in Chapter 6, offered a useful extension of the French and Japanese cases because it allowed us to examine how the existence of domestic resources affected the industrial compacts negotiated between business and government, and what this meant for firm characteristics. US production dominated world markets in the early decades of the oil industry, and this allowed the progeny of Standard Oil to dominate the supply of oil internationally, and to secure advantageous positions within the national markets of the major oil-importing countries. While weakened by the nationalizations of the 1970s, these firms continue to be the most important of the privately owned firms operating in the oil industry. Yet the existence of domestic resources also enabled the existence of a plethora of domestically focused firms that took advantage of smaller opportunities of less interest to the majors. Further, these firms, like those in Japan, continued to prefer the support of the state. The laws and policies used to govern oil markets thus tended to consolidate existing corporate forms.

The case of the United States is also useful because it enables us to examine how the characteristics of firms affected their behavior unencumbered by differences in political institutions in France and Japan that could also plausibly affect their strategies. Firms within the United States were of two main types. The first type emerged as large, vertically integrated oil firms with significant holdings of oil internationally. The second, the independents represented by the Independent Petroleum Association of America (IPAA), were smaller firms and produced almost exclusively within the United States. Here, I found further evidence to support the argument advanced in Chapter 3, although it was also necessary to adjust our expectations because of the existence of significant domestic reserves of oil within the United States. Evidence introduced in Chapter 6 shows that the larger and more diversified US producers were disinterested in retaining protection within the domestic market, as with the French firms Total and Elf Aquitaine. This stood in contrast to smaller producers, which – like their Japanese counterparts – continued to demand support. Industry was united, however, in seeking to increase the opportunities for accessing production with the United States; both types of firm consistently advocating increasing domestic production through the use of trade barriers, or the opening of federal lands or the OCS to greater drilling. Thus the utility of government support to the larger, more diversified firms was greater in the United States than in the other two cases, as a function of the existence of domestic resource endowment.

Moving further afield, in Chapter 2 I showed that governments in a large number of countries made similar changes to the laws and policies governing oil markets. The method of comparative, institutional analysis is useful for tracing the effects of laws and policies on the characteristics of firms, and the way this affected their strategies, including towards government. But how confident can we be that the argument extends to the countries included in the broad survey of changes in oil market governance? On the one hand, the US case gives us greater confidence in the assertions about how firm characteristics affect the strategies they adopt. We can also see that a number of these cases have large and diversified firms – most notably Italy (ENI), Portugal (GALP, formerly Petrogal and others), and Spain (Repsol YPF) – located within domestic borders, while in other countries, most notably Germany, there is no firm of this type. Given the lack of domestic resources (other than Italy), this suggests there is variation across other cases in terms of the characteristics of firms operating in the oil sector that cannot be attributed to differences in resource endowments. The lack of firms with these characteristics in some countries also suggests, on the other hand, that the balance of factors that led to changes in governance in oil markets is likely to vary across cases. Differences in political institutions and the way they aggregate the interests of firms and other actors vary across these cases, and this could affect the ability of firms to get what they want. It is also possible that the position taken by Europe towards the governance of energy markets had a greater effect on other countries than on France because of this, given France's central role within Europe. While the relative importance of these different factors may vary across cases, I have nevertheless presented evidence that suggests both that the strategies of firms are important in explaining the changes in oil market governance discussed, and that the origins of differences between firms can be traced to previous agreements over how to regulate the demand for and supply of oil negotiated between firms and governments.

Broader lessons for the study of economic liberalization

In addition to its intrinsic importance, the case of oil is useful for investigating the changes made to economic governance across sectors in the advanced industrialized states over the past three decades. One reason for this is that it is insulated from a number of important sources of economic liberalization that are external to national economies, most notably the influence of the international trade regime and pressure from the United States. This enables us to better focus on the domestic interest group politics surrounding oil market liberalization without needing to consider how these broader changes external to national economies affect the strategies of domestic actors.

There are three lessons we can take from this study for the broader problem of explaining patterns of economic liberalization. The first focuses on the origins of firm competitiveness. A common approach to the study of trade liberalization uses industries as the unit of analysis, and examines relative factor endowments to explain the competitiveness of industries, and by extension the types of demands they make on governments. More recently, scholars have shown that there are substantial differences between firms operating within single industries, casting doubt on the appropriateness of industry as the best unit of analysis, and on relative factor endowments as the source of firm competitiveness. The case of oil also suggests a more complicated picture of the sources of competitive advantage. The cases examined here show not only that the industrial compacts negotiated between business and government in France and Japan have an important impact on the competitiveness of firms operating in these two countries, but also that resource abundance produced internationally competitive majors and also enabled the emergence of a large number of small but politically powerful oil firms with policy preferences that differ substantially from the majors. This suggests we need to develop better explanations for the origins of the characteristics and competitiveness of firms and industries.

The second lesson focuses on the importance of sequencing in driving economic change. The case of oil shows, in particular, how changes in economic governance today can be understood not only as the result of forces external to domestic political economies, such as the success of the global trade regime in promoting cooperation, falling transportation costs, or pressure from the United States, nor as occurring because of the role of slowing economic growth or shocks in shifting the trajectory of government policy. In the case of oil markets, the loss of control over production by the majors certainly led to a diversification of opportunities internationally and reduced the importance of national markets, even as it increased price volatility. Yet the changes in the governance of oil markets detailed in this book can equally be understood as a result of the way business and government sought in earlier periods to shape the competitive landscape in their favor, and the way this went on to affect the characteristics of firms, and the types of demands they placed on governments.

The role of laws, regulations, and policies in shaping the characteristics of firms is well known to scholars in business studies and comparative political economy. There is also a body of evidence showing that the strategies adopted by firms are conditioned by their characteristics. Yet the crucial relationship between these two views is rarely identified. If we accept that institutions, in addition to factor endowments, can shape firm characteristics, and that firm characteristics affect their strategies towards government, then this book shows that an endogenous explanation for liberalization and protection centered on the firm is possible. This is by no means a necessary outcome. The compact negotiated in Japan, for example, led firms to develop characteristics that led them to continue to invest in retaining existing laws and policies.[1] The likelihood of this occurring is also likely to be contingent on the boundaries of the market the firm is engaged in. It is unsurprising that firms tend to be more willing to invest in protecting the status quo in electricity markets, for example, which remain largely national. Yet it is a truism that many

[1] Douglass C. North, *Institutions, Institutional Change, and Economic Performance* (Cambridge University Press, 1990).

product markets are increasingly defined globally; what determines profitability across a wide range of sectors is the ability to access labor, capital, and markets globally, rather than retaining market share at home. Given this, we should expect that firms and industries that have previously benefitted from the effects of domestic institutions on their competitiveness may over time begin to restructure themselves in order to maximize commercial opportunities abroad. Under these conditions, for firms that have the potential to compete internationally, laws that are the expression of agreements between business and government over how to regulate a market will become increasingly irrelevant at best – or, at worst, will actively hinder – their ability to compete.[2]

The changes made to the industrial compact that governed the Japanese oil markets, despite the opposition of domestic refining firms, point to a third lesson that can be drawn from the study of oil in considering the broader problem of political economic change. Much of the work in the political economy of trade, in particular, focuses on the interests and strategies of producers. The case of oil suggests, however, that the distributional effects of policies on producers and firms that use products as inputs also matter in shaping the interest group politics surrounding economic liberalization. Here oil is an extreme, and therefore a useful, case for exploring the role of competition between lobbies in shaping forms of economic governance.[3] As described in Chapter 5, barriers to trade in oil were abolished in Japan over a ten-year period despite the opposition of domestic refining firms. One reason for this was the poor performance of these firms, but the changing approach of policymakers to governing oil was also influenced by the costs imposed on firms that used oil products as an input into manufacturing. Similarly, in the case of the United States, petrochemical and other firms, as well as

[2] Knight, Hall, and Thelen, among others, emphasize the importance of institutional stability as a political settlement. Peter A. Hall and Kathleen Thelen, "Institutional Change in Varieties of Capitalism," *Socio-Economic Review* 7 (2009): 7–34; Kathleen Thelen, "Institutional Change in Advanced Political Economies," *British Journal of Industrial Relations* 47 (2009): 471–98.

[3] Jason Seawright and John Gerring, "Case Selection Techniques in Case Study Research: A Menu of Qualitative and Quantitative Options," *Political Research Quarterly* 61 (2008): 294–308.

congressional members with important manufacturing centers within their districts, mattered in shaping policy outcomes in ways that mitigated against the resurrection of barriers to trade.

The politics of resource nationalism

The industrial compacts in France, Japan, and the United States were negotiated as firms and governments sought to capture the benefits associated with the production of a product newly important to industrial growth and military power. Today the world oil market is undergoing an extraordinary transformation in the structure of both demand and supply. On the supply side, the United States became a net importer of oil in 1948, and the implications of this have been important in driving oil policy since that time. Yet today the International Energy Agency projects that the United States will become the world's largest oil producer by 2020, and that North America will become a net exporter around 2030, as a result of technological innovation that has cut the costs of extracting unconventional oils.[4] Models of future oil production are often proven faulty. Yet this change, and the possibility that the technologies used to unlock this oil will spread to other countries, has the potential to reorder the structure of supply globally. This suggests that the nature of competition between firms is also likely to change, as is the type of support governments offer them. Furthermore, growing consumption in China and India is shaping the structure of demand globally, and history suggests this is likely to continue for decades into the future regardless of efforts to diversify the energy supply mix and use energy more efficiently. Indeed, the huge populations in these countries and the close relationship between economic growth and energy demand while countries industrialize mean the scale of this demand growth has never been seen before.

What does this study tell us about the consequences of these changes for oil market governance and for international relations? The rise of new industrial powers in the international system – most notably China and India – has introduced a similar set of concerns to those expressed in the past: will this enormous growth in demand lead to

[4] International Energy Agency, *World Energy Outlook 2012* (Paris: OECD/ IEA, 2012), 23.

increasing scarcity, as demand from consumers outstrips the capacity of producers to meet these needs while keeping prices constant? How are these countries likely to reorganize their diplomatic strategies in response to rising demand? What is the relationship between governments and firms in these countries?

Perhaps the most important lesson of this study is that we should not be surprised that firms in the newly industrializing powers are seeking to improve their competitive position and find the support of government useful as they do so. Indeed, the strategies firms and governments in China, India, and elsewhere are using replicate those pursued for decades in other countries. The findings also suggest, however, that the success of firms in competing is likely to reduce, rather than increase the usefulness of government to them. Thus while the diversification of the international oil market described in Chapter 2 offers one reason why we can be sanguine about the long-term implications of the rise of energy use for international relations, a second reason is that the very success of governments and firms in promoting their competitiveness is likely to greater commercialization, rather than politicization, of oil policy in these countries.

This does not mean we can ignore the implications of these changes in demand and supply for international relations. Intervention in oil markets matters because of its distributive effects, transferring wealth from consumers to producers and affecting which fuels are used. The national ownership of substantial reserves of oil in producing countries also means the rise of new consumers has important implications for diplomacy. Subsidies and other forms of support for national oil firms, for example, can lead to side payments from one government to another in the form of military arms, public infrastructure, or other support that functions as an inducement to gain preferential access to production. It can also shift the benefits of exploiting oil from firms based in one country to another. It is also likely to weaken the ability of the United States to impose sanctions on oil-producing states, given its reduced share of aggregate demand in relative terms. If history is a guide, such commercial and diplomatic maneuvering will continue to be a feature of the international oil market, and an important challenge facing governments will lie in managing its repercussions. Yet more than this, the enormous increase in demand for fossil fuels, including oil, in the Asia-Pacific and the Middle East has enormous implications for the global climate. Building cooperative relations with firms and governments in managing not only this new transformation in the energy landscape, but also its environmental implications, is one of the biggest challenges we face in the decades to come.

Appendix: Oil market liberalization in the advanced industrialized states

Country	Ownership	Trade
Austria	In 1980 the federal government owned 100% of OeMU, which handled about 80% of oil production and refining in Austria, and about 66% of oil imports. OeMV also owned and operated the only refinery in Austria, and was active in retail. In 1981 OeMV owned exploration rights to 84% of Austria's prospective areas. In 1987, 15% of OeMV was sold, followed by a further 10% in 1989. By 1992 OeMV was 70% government owned through a holding company. The government reduced its holdings of OeMV to less than 50% by the end of 1994, following a new legal requirement that it had to hold more than 25% but less than 50% of oustanding shares. Austria's refinery was still owned by OeMV in 1996. By the end of 1997, state ownership was reduced to 35%, and in 2007 it was cut to 31.5%.	No "intrinsic" barriers to energy trade, although OeMV dominant as only refinery operator and market power in domestic production and imports. Imports and exports of oil products licensed as of 1983–4. This remained the case in 1989. As of January 1, 1995, oil import licenses were no longer required. This remained in place for both crude and product exports until 1997, when import licenses were abolished and the oil sector opened to full competition.

Country	Ownership	Trade
Belgium	Petrofina, the major firm operating in the petroleum sector with a presence across production and refining, was entirely privately owned in 1983.	Belgian energy markets were free of significant barriers to trade, except in coal. Belgium is an important refining center for Western Europe, and in 1982 more than half the refined products in Belgium were exported.
Finland	Neste Oy was established in 1947 by the government. It operated in refining and retail, and by 1992 also produced in Norway, the UK, US, and elsewhere. In 1995 the company was listed and 16.8% of the firm was sold. Fortum Oyj was created in 1998 in a merger of Neste, Finland's largest oil company, and IVO, its largest electricity supplier. The government owned 61% of Fortum by 2000.	Neste Oy began with the exclusive right to import crude oil, with high dependence on the former Soviet Union. Imports of crude oil and oil products were licensed, with Neste Oy holding a monopoly on crude imports and gasoline. Formal restrictions on imports were lifted in 1990. In July 1991 the government announced that import licenses for oil products were to be granted to all applicants. By January 1, 1992, the system of import licenses for fuels was abolished.
France	In 1992 the government had a shareholding of just over 51% in Elf. The same year it reduced its holdings in Total from 33.87% to 5%. In 1994, the government reduced its shareholding in Elf Aquitaine to 11%, including a "golden share."	Licenses for imports of oil and oil products were formally required until 1993. Formal restrictions on crude imports, and products, were abolished from January 1, 1993.

Country	Ownership	Trade
	In 1998 Elf Aquitaine was fully privatized. Total fell to 5% in 1995 and 0% in 1996. Totalfina and Elf merged in 2000.	
Germany	Deminex was established in 1969 with support of the government, and shareholding from eight German firms. Between 1969 and 1989 it received 2.4 billion deutschemarks in subsidies. Subsidy provision ended at the end of 1989. The government did not take an equity position in the firm, although it was involved in its creation. Refining firms did not receive significant support.	The German oil market was fully liberalized with no import restrictions.
Greece	The Greek Public Petroleum Company (DEP) was controlled by the government and engaged in exploration and production, although it did not have a monopoly on domestic areas from 1980. Two of four oil refineries operating in Greece were government owned. In 1986 the DEP was vertically integrated by reorganizing it into holding company with subsidiaries in E&P, refining, and distribution/marketing, but did not have monopoly status. A new law on E&P was passed in 1994 that treats all firms equally	The government retained a monopoly on imports of crude oil and products, and on distribution of oil products. Crude imports were purchased by the government, usually in accordance with bilateral economic agreements reached with oil-exporting countries. Oil companies could only market and distribute oil products supplied by the government refineries or imported by the government. In 1987 it became unnecessary to obtain a license from

Country	Ownership	Trade
	and removes requirement to conclude two-year contract with domestic refiners. A new hydrocarbons law in 1995 allows private companies to participate in the development and operation of concession areas, and also removes compulsory majority participation of the DEP in concession areas. In 1998 the DEP was merged with other firms to form Hellenic Petroleum, and was transformed into a joint-stock company, of which 23% was sold on the stock exchange. A second tranche was sold in 2000, but the government retained a majority share. By 2001 Hellenic was owned 58.4% by government, 9.6% by a government finance firm, and 32% privately. By 2004 the state holding was 35.4%.	the Ministry of Industry for imports or exports of petroleum. Under new oil law 1571/85 import volumes were fixed at 15% of the total Greek market. This percentage was increased to 25% and then 35% in July, 1987. Imports of oil products continued to be regulated as of 2000.
Ireland	The Irish National Petroleum Corporation (INPC) was established in 1979. It was created under normal company law but shares are owned by the Ministry of Finance. In April 1982, the INPC took control of the County Cork oil refinery. It was required to operate the refinery and supply 55% of national petroleum products	Petroleum prices reflected world market levels and prices were allowed to rise. Importers of petroleum products were required by the 1982 Government Fuels Order to purchase at least 35% of their petrol and gasoil needs from the refinery. In 1984 the European Court of Justice upheld that

Country	Ownership	Trade
	from it. In July 2001 government participation in the oil industry through the Irish National Petroleum Corporation (INPC) ended with the sale of the refinery.	the restrictions on trade were necessary in order to assure the continued operation of the refinery in line with these. The government also stopped requiring importers of oil products to buy at least 35% of their needs from the state-owned refinery, effective in 1997.
Italy	Ente Nazionale Idrocarburi (ENI) and subsidiary AGIP were government owned. It continued to retain preferential rights to certain concession areas domestically in 1992, and the right to acquire gas and oil produced. Under an EU hydrocarbons licensing directive approved in May 1994 Italy was obliged to end ENI's special drilling rights, and remove restrictions on E&P in onshore Sicily. This did not occur until December 1998, when ENI lost its monopoly on E&P in the Po Valley and elsewhere. In 1995, 15% of ENI shares were sold, followed by a further 14% in 1996, and 14% in 1997, reducing government ownership to 51%, 38% in 1998, and 30.8% in 2000.	Export licenses continued to be required in 1983 for the export of crude oil and oil products to non-European Community member countries. Exports of major oil products to non-EC countries required licenses as of 1988. Refineries were banned from importing oil products, and these restrictions remained in place to 1996. By 2003 there were no restrictions on imports, exports or prices.

Country	Ownership	Trade
Japan	The Japan National Oil Corporation (JNOC) was government owned and subsidized private firms' exploration activities overseas and on Japan's continental shelf. It was replaced by the Japan Oil, Gas and Metals National Corporation (JOGMEC) in 2004. JOGMEC is also a government entity subsidizing hydrocarbon and energy exploration.	Imports of oil and oil products were subject to licensing until 1996. In 1985 some firms were allowed to import a wider range of products. Exports of oil products still required government permission. All exports of oil products from Japan were subject to government approval under the Foreign Exchange and Foreign Trade Control Law. From 1996 anyone could import so long as stockpile were maintained.
Netherlands	Integrated oil and gas major Shell is headquartered in the Netherlands. No state ownership.	No barriers to trade. Rotterdam functions as a major European refining center.
Portugal	Petróleos de Portugal S.A. (Petrogal) retained a monopoly on oil refining in 1982. Production within Portugal was open to non-Portuguese firms, but they were required to establish a joint venture with a local firm in 1985. In 1989 Petrogal was changed to a public limited company, to enable private sector participation. Government began to sell 51% of its shares between 1992 and 1995. It sold 25% of the	Oil product imports required authorization from the government. A quota system gave firms operating in the domestic market a monopoly over trade in most fuel products in 1984. In 1986 the quota system in the product market began to be relaxed, and was abolished in 1991.

Country	Ownership	Trade
	firm in 1993. In 1995, it reduced its share to 55%. In 1999, it created a new holding company, GALP Petróleos, with the government holding 60.56% of outstanding shares. In 2000, the new company was named GALP Energia, with the government holding reduced to 34.8%.	
Spain	The National Institute for Hydrocarbons (INH) was established in 1981 to coordinate government holdings, including CAMPSA (retail petroleum sales), ENIEPSA (domestic exploration and production), Petroliber (operation of a refinery in La Coruña), and EMP (operation of three refineries and four major petrochemical subsidiaries). CAMPSA changed its ownership in 1983, with the new share participation 42% by public refiners, 42% by private refiners, and 16% by the state-owned organization INH. The majority of shares after restructuring remained held by the government in a 1984 law. INH owned two other firms doing exploration and production: Hispanoil operated internationally, while ENIEPSA operated within Spain. The two were merged in 1985 as	The government had a monopoly on petroleum sales at retail level since 1927, and controlled oil imports and operated about half the refining industry. A quota system was applied to refiners for imports of crude oil. CAMPSA had the monopoly of distributing petroleum products to private retailers. By decree of December 27, 1985, import restrictions were progressively abolished, with the transition period ending in 1991. Companies of the EC countries were allowed to deliver oil products to Spain under a quota state since 1986. Oil imports from non-EU countries required authorization, as did product exports. The new oil law (Law 34/92) of December 1992 opened the Spanish market to EU

Country	Ownership	Trade
	Hispanoil. In 1986 REPSOL was created, fully owned by the government. In 1989 26% of REPSOL's equity was floated. In 1992 the monopoly of CAMPSA was abolished and the Compañia Logística de Hidrocarburos (CLH) was created, with REPSOL an important shareholder. By 1996 the government retained 10% ownership of REPSOL, along with a "golden share" and a veto over strategic decisions. Privatization was completed in 1998, with the state retaining a "golden share."	and non-EU companies, although licensing was required until 1998.
Sweden	Svenska Petroleum Exploration AB was created in 1969 on the initiative of the Swedish government and a number of Swedish private enterprises for drilling in Sweden, with 50% government ownership. In 1973 Petroswede was created on a 50/50 state/private split to carry out operations internationally. In 1979, Svenska Petroleum purchased Petroswede and changed its name to Svenska Petroleum Exploration. OKP was created in 1986 with minority government ownership. In 1991 the government sold its share in Svenka to OKP, which was 24% government owned. In 1994 OKP was sold.	No mention of quotas, tariffs, or other restrictions for import or export of oil products.

Country	Ownership	Trade
Switzerland	No government ownership.	No quotas, tariffs, or other restrictions for import or export of oil products.
Turkey	In 1980 Turkey's domestic oil was produced by the state oil company TPAO, which controlled 41% of output. In 1981 TPAO was reorganized as a holding company to establish joint ventures with domestic and foreign investors for exploration and development. In 1989 four of five refineries were still owned by the government. TÜPRAŞ, the state refining company, was partially privatized, with the sale of 2.5% of the equity to private investors in May 1991. By 2005 TPAO remained fully government owned, with responsibility for the exploration and production of oil and gas sources both inside and outside the country. TÜPRAŞ, owned 65.8% by the government and 34.2% by the private sector, operates in refining. DITAS is TÜPRAŞ' subsidiary, and transports crude oil to its refineries. The Petroleum Pipeline Company (BOTAŞ), a fully government-owned enterprise, is responsible for oil and natural gas transportation as well as gas importation.	Crude oil imports were monitored by the state-owned refining company TÜPRAŞ, which also had a monopoly on importing oil products. Exporting oil products also required authorization. From 1989 oil companies other than TÜPRAŞ were allowed to carry out oil product imports. Since June 1989 all refineries and retailers were allowed to import oil. Exports of oil products still required government authorization in 1990. Import duties of 14% were levied on crude to protect domestic production in 1996. Oil product imports were liberalized in 1989. Exports were also liberalized but with notification.

Country	Ownership	Trade
	POAS, which has been fully privatized, is the largest distributor and marketer of petroleum products.	

Sources:

International Energy Agency, *Energy Policies and Programmes of IEA Countries* (Paris: OECD/IEA, 1978–2000);

International Energy Agency, *Energy Policies of IEA Countries: Austria* (Paris: OECD/IEA, 1998, 2002);

International Energy Agency, *Energy Policies of IEA Countries: Belgium* (Paris: OECD/IEA, 1997, 2001, 2005);

International Energy Agency, *Energy Policies of IEA Countries: Finland* (Paris: OECD/IEA, 1999, 2003);

International Energy Agency, *Energy Policies of IEA Countries: France* (Paris: OECD/IEA, 1996, 2000);

International Energy Agency, *Energy Policies of IEA Countries: Germany* (Paris: OECD/IEA, 1998, 2002);

International Energy Agency, *Energy Policies of IEA Countries: Greece* (Paris: OECD/IEA, 1998, 2002);

International Energy Agency, *Energy Policies of IEA Countries: Ireland* (Paris: OECD/IEA, 1999, 2003);

International Energy Agency, *Energy Policies of IEA Countries: Italy* (Paris: OECD/IEA, 1999, 2003);

International Energy Agency, *Energy Policies of IEA Countries: Japan* (Paris: OECD/IEA, 1999, 2003);

International Energy Agency, *Energy Policies of IEA Countries: Netherlands* (Paris: OECD/IEA, 1996, 2000, 2004);

International Energy Agency, *Energy Policies of IEA Countries: Portugal* (Paris: OECD/IEA, 1996, 2000, 2004);

International Energy Agency, *Energy Policies of IEA Countries: Spain* (Paris: OECD/IEA, 1996, 2001, 2005);

International Energy Agency, *Energy Policies of IEA Countries: Sweden* (Paris: OECD/IEA, 1996, 2000, 2004);

International Energy Agency, *Energy Policies of IEA Countries: Switzerland* (Paris: OECD/IEA, 1999, 2003);

International Energy Agency, *Energy Policies of IEA Countries: Turkey* (Paris: OECD/IEA, 1997, 2001).

Select bibliography

France

Industry and firm reports

Compagnie Française des Pétroles
Elf Aquitaine
Entreprise de recherches et d'activités pétrolières
Société nationale des pétroles d'Aquitaine
Société nationale elf aquitaine
Total

Official documents

Commission of the European Communities. "The Community Oil Market, Its Oil Refining Industry, and the External Trade in Petroleum Products." COM (86) 263 final. May 13, 1986.

"The Development of an Energy Strategy for the Community: Communication from the Commission to the Council. Mandate of 30 May 1980." COM (81) 540 final. October 2,1981.

"First Guidelines for a Community Energy Policy. Memorandum Presented by the Commission to the Council." *Bulletin of the European Communities*, Supplement to No. 12–1968, COM (68) 1040 final. December 18, 1968.

"Guidelines and Priority Actions under the Community Energy Policy." SEC (73) 1481 final. April 19, 1973.

"The Situation in the Oil Refining Industry and the Impact of Petroleum Product Imports from Third Countries." COM (85) 32 final. March 1, 1985.

Council of the European Communities. "The Internal Energy Market." COM (88) 238 final.

"1171st Meeting of the Council – Energy," June 2, 1987. May 2, 1988.

Court of Justice of the European Union. "Commission of the European Communities vs. French Republic, Judgment of the Court of June 4, 2002." C-483/99. ECR *2002*: I-4781.

Office National des Combustibles Liquides. *Annales de l'Office National des Combustibles Liquides Vol. 1*. Paris: Société d'éditions techniques, 1926.

Special Council of Ministers of the European Coal and Steel Community. "Protocol of Agreement on Energy Problems, Reached Between the Governments of the Member States of the European Communities at the 94th meeting of the Special Council of Ministers of the European Coal and Steel Community." *Official Journal of the European Union*. April 21, 1964: 1099–1100.

Periodicals

International Petroleum Finance
Le Figaro
Le Monde
Oil & Gas Journal
Petroleum Economist
Petroleum Intelligence Weekly

Japan

Industry and firm annual reports

Idemitstu Kosan. *Idemitsu Sekiyu Shiryo Showa [Idemitsu Petroleum Materials]*. Tokyo: Idemitsu, various years.

Japan Oil, Gas and Metals National Corporation. *Heisei 17-Nendo Keikaku [Corporate Plan for 2005]* (Tokyo: JOGMEC, 2004.

Keidanren. "Sekiyu Wo Chushin to Suru Enerugii ni Kansuru Yobo [Request in Relation to Petroleum and Other Energy Resources]." Tokyo: Keidanren, 1973.

Sekiyu Kogyo Renmei. *Wagakuni Sekiyu Kaihatsu no Genjo [The State of Japan's Petroleum Development]* Tokyo: Sekiyu Kogyo Renmei, various years.

"*Wagakuni Sekiyu/Tennen Gasu Kaihatsu No Genjo to Mondaiten [Current State and Problems in Japan's Petroleum and Natural Gas Development]*." Tokyo: Sekiyu Kogyo Renmei, various years.

Sekiyu Renmei. "*Sekiyu Gyokai No Suii [Trends in the Petroleum Industry]*." Tokyo: Sekiyu Renmei, various years.

Sekiyu/Enerugii Seisaku ni Kan Suru Teigen [Proposal Regarding Petroleum and Energy Policy]. Tokyo: Sekiyu Renmei, 1982.

Official documents

Ministry of International Trade and Industry. "Kongo no Sekiyu Seisaku no Hoko [The Current and Future Direction of Petroleum Policy]." Tokyo: Ministry of International Trade and Industry, 1978.

"Kongo no Sekiyu Sangyo no Arikata ni tsuite (Sanko Shiryo) [Internal Report: On the Future of the Petroleum Industry: Additional Material]." Tokyo: Ministry of International Trade and Industry, 1981.

"Motouri Kigyo no Shuyakuka ni Tsuite Chukan Hokoku [Interim Report on the Concentration of Wholesale Firms]." Tokyo: Ministry of International Trade and Industry, 1984.

"Sekiyu Ryutsu Bijon Kenkyukai – Chukan Hokoku: Shiryo 1: Shohisha Ni-zu ni Kansuru Giron Pointo [Interim Report of the Research Committee for a Vision for Petroleum Distribution: Materials 1: Summary of Debate on Consumer Needs]." Tokyo: Ministry of International Trade and Industry, 1984.

"Sekiyu Sangyo no Kozo Kaizen No Hoko to Sekiyu Sangyo no Arikata ni Tsuite [On the Direction of Structural Reform in the Petroleum Industry and the State of the Petroleum Industry]." Tokyo: Ministry of International Trade and Industry, 1985.

"Sekiyu Sangyo ni okeru Kozo Kaizen no Genjo to Kongo no Hoko [The Current State and Future Direction of Structural Reform in the Petroleum Industry]." Tokyo: Ministry of International Trade and Industry, 1983.

Sekiyu Bukai. "Sekiyu Sangyo no Keiri Jokyo to Seihin Kakaku Mondai [On Problems Related to the Financial Position of the Petroleum Industry and Petroleum Product Prices]." Tokyo: Ministry of International Trade and Industry, 1981.

"Naibu Shiryo: Kongo No Sekiyu Sangyo No Arikata ni Tsuite [Internal Report: On the Future of the Petroleum Industry]." Tokyo: Ministry of International Trade and Industry, 1981.

Sekiyu Kihon Mondai Kenkyukai. "Sekiyu Kihon Mondai Kenkyukai Chukan Hokokusho [Interim Report of the Research Committee on Fundamental Problems in Petroleum]." Tokyo: Ministry of International Trade and Industry, 1981.

Sekiyu Sangyo Kihon Mondai Kentokai. "Sekiyu Sangyo Kihon Mondai Kentokai Hokokusho [Meeting Minutes of Report of the Deliberative Committee into the Petroleum Industry]." Tokyo: Ministry of International Trade and Industry, 1988.

Sekiyu Shingikai Sekiyu Bukai Shoiinkai. "Naibu Shiryo: Hiaringu ni Okeru Kakushabetsu Iken [Internal Memorandum: Firm Positions in Hearing]." Tokyo: Ministry of International Trade and Industry, 1981.

"Hiaringu Kaitosho [Hearing Response Form]." Tokyo: Ministry of International Trade and Industry, May 30, 1985.

"Hiaringu Kaitosho [Hearing Response Form]." Tokyo: Ministry of International Trade and Industry, June 3, 1985.

Shigen Kenkyu Iinkai. "Hyo 8–3: Wagakuni no Chiiki Betsu, Shigen Betsu, Kaigai Kaihatsu Projekuto [Graph 8–3: Japan's Overseas Development Projects by Region and Resource-type]. In *Shigen Kenkyu Iinkai Kyoryoku Kaihatsu Mondai Senmon Iinkai: Shingi Ronten [Natural Resource Research Committee Cooperative Development Problems Specialist Committee: Debate Summary]*. Tokyo: Ministry of International Trade and Industry, November 24, 1971.

"Hyo A-3: Kaigai Sekiyu Kaihatsu Jigyo Ichiran [Graph A-3: Overseas Petroleum Development Projects]." In *Shigen Kenkyu Iinkai Kyoryoku Kaihatsu Mondai Senmon Iinkai: Shingi Ronten [Natural Resource Research Committee Cooperative Development Problems Specialist Committee: Debate Summary]*. Tokyo: Ministry of International Trade and Industry, November 24, 1971.

Sogo Enerugi Chosakai. "Kongo no Sekiyu Sangyo no Arikata ni Tsuite (Sanko Shiryo) [Internal Report: On the Future of the Petroleum Industry: Additional Material]." Tokyo: Ministry of International Trade and Industry, 1981.

"Statement by Kaoru Yosano, Minister for International Trade and Industry," on October 13, 1998, to the Economy and Industry Committee, Japanese House of Representatives.

"Testimony of Keijiro Murata, Minister of International Trade and Industry," on November 20, 1985, to the Committee on Commerce and Industry, Japanese House of Representatives.

"Testimony of Takashi Nonouchi, Director General of Agency of Natural Resources and Energy," on September 3, 1985, to the Committee on Commerce and Industry, Japanese House of Representatives.

"Testimony of Yasuo Fukuoka, on November 20, 1985, to the Trade and Commerce Committee, Japanese House of Representatives.

Periodicals

Asahi Shinbun
Shukan Toyo Keizai
Nihon Keizai Shinbun

United States

Industry and firm annual reports

Chevron
ChevronTexaco
Conoco

Exxon
Exxon Mobil
Independent Petroleum Association of America. *IPAA Oil and Gas Producing Industry in Your State*. Washington, DC: IPAA, various years.
Mobil
Standard Oil (New Jersey)
Texaco

Official documents

Commodity Futures Trading Commission. *US Interagency Taskforce on Commodities Markets, Interim Report on Crude Oil*. Washington DC: CFTC, 2008.

Department of the Interior. *Report of the Commissioner of Corporations on the Petroleum Industry Vol. 1: Position of the Standard Oil Company in the Petroleum Industry*. Washington, DC: Government Printing Offices, 1907.

Report of the Commissioner of Corporations on the Petroleum Industry Volume II: Prices and Profits. Washington DC: Government Printing Offices, 1907.

Department of State. "Message from the President of the United States Transmitting in Response to Senate Resolution No. 149 of February 13, 1924: A Report of the Secretary of State Relative to the Diplomatic Correspondence in Connection with Securing Oil Concessions for American Citizens between this Government and the Governments of Certain Foreign Countries Regarding Oil Concessions in those Countries." Washington, DC: Government Printing Office, 1924.

"Informal Memorandum." American Embassy in Japan to the Japanese Ministry of Foreign Affairs, 893.6363 Manchuria 29, in *FRUS Japan: 1931–1941 Volume I*.

"Oral Statements by the American Ambassador in Japan (Grew) to the Japanese Minister for Foreign Affairs (Hirota)." 893.6363 Manchuria/194, in *FRUS Japan: 1931–1941 Volume I*.

Senate Committee on Foreign Relations. *A Documentary History of the Petroleum Reserves Corporation 1943–1944*. Washington, DC: US Government Printing Office, 1974.

Senate Special Committee Investigating Petroleum Resources. *American Petroleum Interests in Foreign Countries*, 79th Congress, 1st session, 1945.

Hearings

Committee on Energy and Commerce, United States House of Representatives
Committee on Interior and Insular Affairs, United States House of Representatives

Committee on Interstate and Foreign Commerce, United States House of Representatives
Committee on Resources, United States House of Representatives
Committee on Ways and Means, United States House of Representatives
Committee on Energy and Natural Resources, United States Senate
Committee on Finance, United States Senate
Committee on the Judiciary, United States Senate

Periodicals

New York Times
Oil & Gas Journal
Washington Post

Interviews

France

Olivier Abadie: Crude and Products Trader, Elf Aquitaine (1990–5)
Olivier Appert: Director of Hydrocarbons (1989–93), Ministry of Industry
Gilles Bellec: Director of Hydrocarbons (1984–9), Ministry of Industry
Rene Chappaz: Chief Executive Officer, Total USA
Gilles Darmois: Counselor to Martin Malvy, former State Secretary to the Minister of Industrial Redeployment and Foreign Trade, responsible for energy (1984–6); Advisor to the President of Elf Aquitaine (1991–4)
Bertrand Deroubaix: Vice-President, Strategy and Business Intelligence Division, Total
Dominique Finon: Senior Research Fellow, French National Center of Scientific Research
Maurice Girault: Statistical Service, Department of Ecology, Sustainable Development, Transportation and Housing
Jean-Daniel Levi: Energy Advisor to President Mitterrand (1982–8)
Claude Mandil: Office of Prime Minister (1981–3); Directeur General Énergie et Material Premierés, Ministry of Economy (1990–8); Executive Director, International Energy Agency (2003–7)
Xavier Preel: General Manager, Total Exploration and Production, Total (1991–1996)
Jean du Rusquec: Advisor to the CEO, Total
Bruno Weymuller: Executive Vice-President, Total (2000–8)

Japan

Mitsuo Arai: Committee Member, Petroleum Council (2001–7)

Hiroyuki Fukano: Director of Upstream Division (1998–2000), Ministry of Economy, Trade and Industry

Nobuo Hata: General Manager, Planning and General Coordination Department, Petroleum Association of Japan

Kazutomo Irie: Petroleum Division (1985–6), Ministry of Economy, Trade and Industry

Akira Ishii: Chief Economist, Japan Oil, Gas and Metals National Corporation

Kazuo Kawahara: Director, Petroleum Association of Japan

Takeo Kikkawa: Committee Member, Petroleum Council (2001–)

Shigeru Kikuchi: Deputy Manager, Overseas Operation Department, Idemitsu Kosan

Nobuyori Kodaira: Director, Petroleum Planning Division (1995–6); Administrative Vice Minister, Agency for Natural Resources and Energy (2004–6), Ministry of Economy, Trade and Industry

Hiroshi Kubota: Deputy Director General Oil and Gas Project Development Group, Japan Oil, Gas and Metals National Corporation

Kentaro Morita: Deputy Director Upstream Division (2000–2), Ministry of Economy, Trade and Industry

Masahisa Naito: Director, Petroleum Planning Division (1985–6), Ministry of Economy, Trade and Industry

Taro Shoji: General Manager, Planning and Research Department, Japan Petroleum Development Association

Shigeru Sudo: Former General Manager, Planning and General Coordination Department, Petroleum Association of Japan

Katsuhiko Suetsugu: Member, Petroleum Council, Ministry of Economy, Trade and Industry

Toshiaki Tanaka: Mitsubishi Oil and JOGMEC Technical Committee

Tsutomu Toichi: Managing Director, Institute of Energy Economics Japan; Member, Petroleum Council, Ministry of Economy, Trade and Industry

United States

Guy Caruso: International Energy Agency (1993–8); US Energy Information Administration, Department of Energy (2002–8)

Leonard L. Coburn: Director of the Office of Competition, Department of Energy (1972–92); Director of the Office of Oil Policy, Department of Energy (1992–6)

Charlie Cooke: Professional Staff Member, United States House Committee on Science and Technology

Charles Ebinger: Federal Energy Administration (1975–6)

Alan Hegburg: Deputy Assistant Secretary for International Energy Policy, Department of Energy; Director of International and Geopolitical Analysis, Phillips Petroleum

Bob McNally: Advisor, National Security Council and National Economic Council (2001–3)

John Northington: Special Assistant to Assistant Secretary of the Interior for Land and Minerals, Department of the Interior (1998–2000)

Secondary sources

Abdelal, Rawi. *Capital Rules: The Construction of Global Finance*. Cambridge, MA: Harvard University Press, 2007.

Adams, William J. and Christian Stoffaës, eds. *French Industrial Policy*. Washington, DC: The Brookings Institution, 1986.

Adcock, Robert and Paul Collier. "Measurement Validity: A Shared Standard for Qualitative and Quantitative Research." *American Political Science Review* 95 (2001): 529–46.

Adelman, Morris A. *The Economics of Petroleum Supply: Papers by M.A. Adelman, 1962–1993*. Cambridge, MA: MIT Press, 1993.

The Genie out of the Bottle: World Oil since 1970. Cambridge, MA: MIT Press, 1995.

"Mineral Depletion, with Special Reference to Petroleum." *The Review of Economics and Statistics* 72 (1990): 1–10.

"Scarcity and World Oil Prices." *The Review of Economics and Statistics* 68 (1986): 387–97.

The World Oil Market to 1970. Cambridge, MA: Massachusetts Institute of Technology Center for Energy Policy Research, 1988.

The World Petroleum Market. Washington, DC: Resources for the Future, 1972.

el-Agraa, Ali. M. and Yao Suhu. "National Versus Supranational Interests and the Problem of Establishing an Effective EC Energy Policy." *Journal of Common Market Studies* 22 (1984): 333–49.

Alhajji, A.F. and David Huettner. "OPEC and Other Commodity Cartels: A Comparison." *Energy Policy* 28 (2000): 1151–64.

Alt, James E. and Michael Gilligan. "The Political Economy of Trading States: Factor Specificity, Collective Action Problems and Domestic Political Institutions." *Journal of Political Philosophy* 2 (1994): 165–92.

Alt, James E., Fredrik Carlsen, Per Heum, and Kåre Johansen. "Asset Specificity and the Political Behavior of Firms: Lobbying for Subsidies in Norway." *International Organization* 53 (2003): 99–116.

Alt, James E., Jeffry Frieden, Michael Gilligan, Dani Rodrik, and Ronald Rogowski. "The Political Economy of International Trade: Enduring Puzzles and an Agenda for Inquiry." *Comparative Political Studies* 29 (1996): 689–717.

American Petroleum Institute. *American Petroleum Industry: A Survey of the Present Position of the Petroleum Industry and Its Outlook toward the Future*. New York: American Petroleum Institute, 1935.

Petroleum; the Story of an American Industry. New York: American Petroleum Institute, 1949.

Andersen, Svein S. "EU Energy Policy: Interest Interaction and Supranational Authority." *ARENA Working Papers* WP 00/5 (2000): 1–23.

Anderson, Irvine H., Jr. *The Standard-Vacuum Oil Company and United States East Asian Policy 1933–1941*. Princeton University Press, 1975.

Anderson, Mark W. "Regulatory Policy toward Domestic Crude Oil Production." *Policy Sciences* 12 (1980): 245–64.

Antill, Nick and Robert Arnott. "Creating Value in the Oil Industry." *Institutional and Corporate Finance* 16 (2005): 18–31.

Oil Company Crisis: Managing Structure, Profitability, and Growth. Oxford Institute of Energy, 2002.

Bacon, Raymond Foss and William A. Hamor. *The American Petroleum Industry*. New York: McGraw-Hill, 1916.

Bahgat, Gawdat. *American Oil Diplomacy in the Persian Gulf and the Caspian Sea*. Gainesville: University Press of Florida, 2003.

Baldwin, Robert. *The Political Economy of US Trade Policy*. Cambridge, MA: MIT Press, 1985.

Bamberg, J.H. *The History of the British Petroleum Company, Volume 2: The Anglo-Iranian Years, 1928–1954*. Cambridge University Press, 1994.

Bantekas, Ilias. "Hellenic (Greek) Petroleum Law: Regulation of Exploration and Exploitation." *Journal of Energy and Natural Resources Law* 12 (2002): 117–29.

Barber, William J. "The Eisenhower Energy Policy: Reluctant Intervention." In *Energy Policy in Perspective: Today's Problems, Yesterday's Solutions*, edited by Craufurd D.W. Goodwin, 205–86. Washington, DC: Brookings Institution, 1981.

"Studied Inaction in the Kennedy Years." In *Energy Policy in Perspective: Today's Problems, Yesterday's Solutions*, edited by Craufurd D.W. Goodwin, 287–336. Washington, DC: Brookings Institution, 1981.

Baron, David. "Integrated Strategy, Trade Policy, and Global Competition." *California Management Review* 39 (1997): 145–69.

Barsky, Robert B. and Lutz Kilian. "Oil and the Macroeconomy since the 1970s." *The Journal of Economic Perspectives* 18 (2004): 115–34.

Beach, Derek. *The Dynamics of European Integration: Why and When EU Institutions Matter*. Basingstoke and New York: Palgrave Macmillan, 2005.

Bellanger, Serge and Olivier Blanchard. "Economic Policy." In *Transformations in French Business: Political, Economic, and Cultural Changes from 1981–1987*, edited by Judith Frommer and Janice McCormick, 35–50. Westport, CT: Greenwood Press, 1989.

Beltran, Alain. *Elf Aquitaine: From its Origins to the Eighties*. Paris: Elf Aquitaine, 1999.

Berkma, Pelin, Sam Ouliaris, and Hossein Samiei. *The Structure of the Oil Market and Causes of High Prices*. New York: International Monetary Fund, 2005.

Bernard, Andrew B., J. Bradford Jensen, Stephen J. Redding, and Peter K. Schott. "Firms in International Trade." *NBER Working Paper No. 13054* (April 2007).

Berne, Michel and Gerard Pogorel. "Privatization Experiences in France." *CESifo Working Paper* (2004).

Blair, John M. *The Control of Oil*. New York: Vintage, 1978.

Blanchard, Olivier J. and Jordi Galí. "The Macroeconomic Effects of Oil Price Shocks: Why are the 2000s so Different from the 1970s?" *NBER Working Paper No. W13368* (September 2007).

Blyth, Mark. *Great Transformations: Economic Ideas and Institutional Change in the Twentieth Century*. Cambridge University Press, 2002.

Boddewyn, Jean J. and Thomas L. Brewer. "International-business Political Behavior: New Theoretical Directions." *Academy of Management Review* 19 (1994): 119–43.

Bohi, Douglas R. and Milton Russell. *Limiting Oil Imports: An Economic History and Analysis*. Baltimore, MD: Johns Hopkins University Press, 1978.

Bohi, Douglas R., Michael A. Toman, and Margaret A. Walls. *The Economics of Energy Security*. Boston, MA: Kluwer Academic Publishers, 1996.

Boix, Carles. "Partisan Governments, the International Economy, and Macroeconomic Policies in Advanced Nations, 1960–93." *World Politics* 53 (2000): 38–73.

"Privatizing the Public Business Sector in the Eighties: Economic Performance, Partisan Responses and Divided Government." *British Journal of Political Science* 27 (1997): 473–96.

Brabant, Stéphane A. and Bertrand Montembault-Héveline. "Reform of French Mineral Law." *Journal of Energy and Natural Resources Law* 15 (1997): 248–62.

Brooks, Sarah M. "Interdependent and Domestic Foundations of Policy Change: The Diffusion of Pension Privatization around the World." *International Studies Quarterly* 49 (2005): 273–94.

Burrows, James C. and Thomas A. Domencich. *An Analysis of the United States Oil Import Quota*. Lexington, MA: Heath Lexington Books, 1970.

Cabinet Task Force on Oil Import Control. *The Oil Import Question*. Washington, DC: Cabinet Task Force on Oil Import Control, 1970.

Campbell, John L. and Ove K. Pedersen. "The Rise of Neoliberalism and Institutional Analysis." In *The Rise of Neoliberalism and Institutional Analysis*, edited by John L. Campbell and Ove K. Pedersen, 1–24. Princeton University Press, 2001.

Canes-Wrone, Brandice. *Who Leads Whom? Presidents, Policy, and the Public*. University of Chicago Press, 2006.

al-Chalabi, Fadhil J. *The World Oil Price Collapse of 1986: Causes and Implications for the Future of OPEC*. Washington, DC: Johns Hopkins University Press, 1987.

de Carmoy, Guy. "The European Alternatives: The Community Energy Policy: Oil." *Government and Opposition* 12 (1977): 411–29.

Chandler, Alfred D., Jr. *Strategy and Structure: Chapters in the History of the American Industrial Enterprise*. Cambridge, MA: MIT Press, 1969 [1962].

Ching, Chih Chen. "Crude Oil Prices and the Postwar Japanese Refining Industry." PhD diss., Massachusetts Institute of Technology, 1967.

Chubb, John E. *Interest Groups and the Bureaucracy: The Politics of Energy*. Stanford University Press, 1983.

Clo, Alberto. *Oil Economics and Policy*. Dordrecht: Kluwer Academic Publishers, 2000.

Cohen, Elie. "Industrial Policies in France: The Old and the New." *Journal of Industry, Competition and Trade* 7 (2007): 213–27.

Cohen, Jeffrey E. "Presidential Rhetoric and the Public Agenda." *American Journal of Political Science* 39 (1995): 87–107.

Cowhey, Peter F. *The Problems of Plenty: Energy Policy and International Politics*. Berkeley, CA: University of California Press, 1985.

Cox, Gary W. and Mathew D. McCubbins. *Legislative Leviathan: Party Government in the House*, 2nd edn. Cambridge University Press, 2007.

Croome, John. *Reshaping the World Trading System: A History of the Uruguay Round*. The Hague: Kluwer Law International, 1999.

Crouch, Colin and Wolfgang Streeck. "Introduction: The Future of Capitalist Diversity." In *Political Economy of Modern Capitalism: Mapping Convergence and Diversity*, edited by Colin Crouch and Wolfgang Streeck, 1–18. London: Sage, 1997.

Culpepper, Pepper D. "Capitalism, Coordination, and Economic Exchange: The French Political Economy since 1985." In *Changing France: The Politics that Markets Make*, edited by Pepper D. Culpepper, Peter A. Hall and Bruno Palier, 29–49. New York: Palgrave Macmillan, 2006.

Cyert, Richard M. and James G. March. *A Behavioral Theory of the Firm*. Malden, MA: Blackwell, 1992 [1963].

Dafter, Ray. "World Oil Production and Security of Supplies." *International Security* 4 (1979): 154–76.

Dahl, Carol and Mine Yücel. "Testing Alternative Hypotheses of Oil Producer Behavior." *The Energy Journal* 12 (1991): 117–38.

Dam, Kenneth W. "Implementation of Import Quotas." *Journal of Law and Economics* 14 (1971): 1–60.

Darden, Keith A. *Economic Liberalism and its Rivals: The Formation of International Institutions among the Post-Soviet States*. Cambridge University Press, 2009.

Davenport, E.H. and Sidney Russell Cooke. *The Oil Trusts and Anglo-American Relations*. London: Macmillan, 1923.

Davis, Christina. *Food Fights over Free Trade: How International Institutions Promote Agricultural Trade Liberalization*. Princeton University Press, 2003.

Davis, David Howard. *Energy Politics*, 3rd edn. New York: St. Martin's Press, 1982.

Davis, Jerome D. *The Changing World of Oil: An Analysis of Corporate Change and Adaptation*. Aldershot: Ashgate Publishing, 2006.

DeGolyer and MacNaughton. *Twentieth Century Petroleum Statistics: Historical Data*. Texas: DeGolyer and MacNaughton, 2000.

DeNovo, John A. "The Movement for an Aggressive American Oil Policy Abroad, 1918–1920." *The American Historical Review* 61 (1956): 854–76.

Desta, Melaku Geboye. "The GATT/WTO System and International Trade in Petroleum: An Overview." *Journal of Energy and Natural Resources Law* 21 (2003): 385–98.

"The Organization of Petroleum Exporting Countries, the World Trade Organization, and Regional Trade Agreements." *Journal of World Trade* 37 (2003): 523–51.

Devarajan, Shantayanan and Anthony C. Fisher. "Hotelling's 'Economics of Exhaustible Resources': Fifty Years Later." *Journal of Economic Literature* 19 (1981): 65–73.

Dobbin, Frank. *Forging Industrial Policy: The United States, Britain, and France in the Railway Age*. Cambridge University Press, 1994.

Downs, Erica S. "The Fact and Fiction of Sino-African Energy Relations." *China Security* 3 (2007): 42–68.

Drezner, Daniel W. *All Politics is Global: Explaining International Regulatory Regimes*. Princeton University Press, 2008.

Dumontier, Pascal and Claude Laurin. "The Financial Impact of the French Government's Nationalization/Privatization Strategy." *EFMA 2003 Helsinki Meetings* (2003).

Edwards, George C., William Mitchell, and Reed Welch. "Explaining Presidential Approval: The Significance of Issue Salience." *American Journal of Political Science* 39 (1995): 108–34.

Engerman, Stanley L. and Kenneth L. Sokoloff. "Factor Endowments, Institutions, and Differential Paths of Growth Among New World Economies: A View from Economic Historians of the United States." *NBER Working Paper Series on Historical Factors in Long Run Growth* 66 (1994).

Estrin, Saul and Peter Holmes. "French Planning and Industrial Policy." *Journal of Public Policy* 3 (1983): 131–48.

Fattouh, Bassam. "An Anatomy of the Crude Oil Pricing System." *Oxford Institute for Energy Studies WPM* 40 (January 2011).

Federal Trade Commission. *The International Petroleum Cartel – Staff Report to the Federal Trade Commission*. Washington, DC: United States Government Printing Office, 1952.

Feigenbaum, Harvey B. *The Politics of Public Enterprise: Oil and the French State*. Princeton University Press, 1985.

"States, Markets and the Politics of Energy." *Polity* 20 (1987): 167–78.

Finon, Dominique. "French Energy Policy: The Effectiveness and Limitations of Colbertism." In *European Energy Policies in a Changing Environment*, edited by Francis McGowan, 21–56. Heidelberg, Germany: Physica Verlag, 1996.

Fitzgerald, Edward Peter. "France's Middle Eastern Ambitions, the Sykes-Picot Negotiations, and the Oil Fields of Mosul, 1915–1918." *The Journal of Modern History* 66 (1994): 697–725.

Foreign Policy Research Institute. *Oil Diplomacy: The Atlantic Nations in the Oil Crisis of 1978–79*. Philadelphia, PA: Foreign Policy Research Institute, 1980.

Frank, Helmut Jack. *Crude Oil Prices in the Middle East: A Study in Oligopolistic Price Behavior*. New York: Praeger, 1966.

Frankel, Paul H. *Essentials of Petroleum: A Key to Oil Economics*. New York: Frank Cass, 1983.

Frieden, Jeffry. *Debt, Development, and Democracy*. Princeton University Press, 1991.

Frieden, Jeffry and Ronald Rogowski. "The Impact of the International Economy on National Policies: An Analytical Overview." In *Internationalization and Domestic Politics*, edited by Robert O. Keohane and Helen V. Milner, 25–47. New York: Cambridge University Press, 1996.

Fusaro, Peter C. "The Future of Oil: Geopolitical Lynchpin or Common Commodity?" In *Global Markets and National Interests: The New Geopolitics of Energy, Capital, and Information*, edited by Lincoln Palmer Bloomfield, 13–57. Washington, DC: CSIS Press, 2002.

Garrett, Geoffrey and Barry R. Weingast. "Ideas, Interests, and Institutions." In *Ideas and Foreign Policy: Beliefs, Institutions, and Political Change*, edited by Judith Goldstein and Robert O. Keohane, 173–206. Ithaca: Cornell University Press, 1993.

Gately, Dermot. "Strategies for OPEC's Pricing and Output Decisions." *Energy Journal* 16 (1995): 1–38.

"A Ten-Year Retrospective: OPEC and the World Oil Market." *Journal of Economic Literature* 22 (1984): 1100–14.

Gawande, Kishgore, Pravin Krishna, and Marcelo Olarreaga. "Lobbying Competition over Trade Policy." *NBER Working Paper 11371* (2005).

Ghosh, Arabinda. *Competition and Diversification in the United States Petroleum Industry*. Westport, CT: Quorum Books, 1985.

Giraud, Andre L. "Energy in France." *Annual Review of Energy* 8 (1983): 165–91.

Goldstein, Judith. "International Institutions and Domestic Politics: GATT, WTO and the Liberalization of International Trade." In *The WTO as an International Organization*, edited by Anne O. Kreuger, 133–52. University of Chicago Press, 1998.

Goodman, John B. and Louis W. Pauly. "The Obsolescence of Capital Controls? Economic Management in an Age of Global Markets." *World Politics* 46 (1993): 50–82.

Gray, Lewis Cecil. "Rent under the Assumption of Exhaustibility." *The Quarterly Journal of Economics* 28 (1914): 466–89.

Grayson, Leslie E. *National Oil Companies*. Chichester: John Wiley and Sons, 1981.

Grossman, Gene M. and Elhanan Helpman. "Protection for Sale." *The American Economic Review* 84 (1994): 833–50.

Guyomarch, Alain. "The Europeanization of Policy-Making." In *Developments in French Politics 2*, edited by Alain Guyomarch, Peter A. Hall, Jack Hayward, and Howard Machin, 116–36. London: Palgrave Macmillan, 2001.

Hacker Jacob S. *The Divided Welfare State*. Cambridge University Press, 2002.

"Privatizing Risk without Privatizing the Welfare State: The Hidden Politics of Social Policy Retrenchment in the United States." *American Political Science Review* 98 (2004): 243–60.

Haigh, Robert. *The Growth of Integrated Companies*. Norwood, MA: Plimpton Press, 1954.

Hall, Peter A. *Governing the Economy: The Politics of State Intervention in Britain and France*. New York: Oxford University Press, 1986.
"Policy Paradigms, Social Learning, and the State: The Case of Economic Policymaking in Britain." *Comparative Politics* 25 (1993): 275–96.
Hall, Peter A. and David Soskice. "An Introduction to the Varieties of Capitalism." In *Varieties of Capitalism: The Institutional Foundations of Comparative Advantage*, edited by Peter A. Hall and David Soskice, 1–68. Oxford University Press, 2001.
Hall, Peter A. and Kathleen Thelen. "Institutional Change in Varieties of Capitalism." *Socio-Economic Review* 7 (2009): 7–34.
Hamilton, James D. "Oil and the Macroeconomy since World War II." *The Journal of Political Economy* 91 (1983): 228–48.
Hancke, Bob. *Large Firms and Institutional Change: Industrial Renewal and Economic Restructuring in France*. Oxford University Press, 2002.
Hartshorn, J.E. *Politics and World Oil Economics: An Account of the International Oil Industry in its Political Environment*, revised edn. New York: Frederick A. Praeger, 1967.
Hein, Laura Elizabeth. *Fueling Growth: The Energy Revolution and Economic Policy in Postwar Japan*. Cambridge, MA: Harvard University Press, 1990.
Helleiner, Eric. *States and the Reemergence of Global Finance: From Bretton Woods to the 1990s*. Ithaca: Cornell University Press, 1994.
Hillman, Arye L. "Declining Industries and Political-Support Protectionist Motives." *American Economic Review* 72 (1982): 1180–7.
Hirschman, Albert O. *National Power and the Structure of Foreign Trade*. Los Angeles: University of California Press, 1980 [1945].
Hiscox, Michael J. "Commerce, Coalitions, and Factor Mobility: Evidence from Congressional Votes on Trade Legislation." *American Political Science Review* 96 (2002): 593–608.
Horsnell, Paul and Robert Mabro. *Oil Markets and Prices: The Brent Market and the Formation of World Oil Prices*. Oxford University Press, 1993.
Hotelling, Harold. "The Economics of Exhaustible Resources." *The Journal of Political Economy* 39 (1931): 137–75.
Hubbert, M.K. "Energy Resources: A Report to the Committee on Natural Resources." National Academy of Sciences, Government Printing Office, Publication No. 1000D (1962).
Hughes, Llewelyn and Francisco Flores-Macias. "Drill Baby Drill! The Correlates of 'Energy Independence' Policy in the United States." Institute for International Economic Policy, George Washington University Working Paper IIEP-WP-2012–4 (2012).

Hughes, Llewelyn and Phillip Lipscy. "The Politics of Energy." *Annual Review of Political Science* 16 (2013): 449–69.

Huntington, Hillard G. "Oil Price Forecasting in the 1980s: What Went Wrong?" *Energy Journal* 15 (1994): 1–23.

Iguchi, Tosuke. *Nihon Sangyo Hattatsushi: Sekiyu [The History of Industrial Development in Japan: Petroleum]*. Tokyo: Gendai Nihon Sangyo Hattatsushi Kenkyukai, 1963.

Ikenberry, G. John. "History's Heavy Hand: Institutions and the Politics of the State." Paper prepared for a conference on "New Perspectives on Institutions," University of Maryland, October 1994.

"The Irony of State Strength: Comparative Responses to the Oil Shocks in the 1970s." *International Organization* 40 (1986): 105–37.

"Market Solutions for State Problems: The International and Domestic Politics of American Oil Decontrol." *International Organization* 42 (1988): 151–77.

Reasons of State : Oil Politics and the Capacities of American Government, Cornell Studies in Political Economy. Ithaca: Cornell University Press, 1988.

International Energy Agency. *The Role of IEA Governments in Energy: 1996 Update*. Paris: International Energy Agency, 1996.

World Energy Outlook. Paris: OECD/IEA, various years.

Ise, John. *United States Oil Policy*. New Haven: Yale University Press, 1928.

Jabko, Nicholas. *Playing the Market: A Political Strategy for Uniting Europe, 1985–2005*. Ithaca: Cornell University Press, 2005.

Johnson, Chalmers A. *MITI and the Japanese Miracle: The Growth of Industrial Policy, 1925–1975*. Stanford University Press, 1982.

Johnson, Simon and James Kwak. *13 Bankers: The Wall Street Takeover and the Next Financial Meltdown*. New York: Vintage, 2011.

Jones, Charles O. and Randall Strahan. "The Effect of Energy Politics on Congressional and Executive Organization in the 1970s." *Legislative Studies Quarterly* 10 (1985): 151–79.

Jones, Donald W., Paul N. Leiby, and Inja K. Paik. "Oil Price Shocks and the Macroeconomy: What Has Been Learned since 1996." *The Energy Journal* 25 (2004): 1–32.

Kalt, Joseph P. "The Creation, Growth, and Entrenchment of Special Interests in Oil Price Policy." *Harvard Institution of Economic Research Discussion Paper Series* 939 (1982): 1–33.

Kaminsky, Graciela and Sergio Schmukler. "Short-Run Pain, Long-Run Gain: Financial Liberalization and Stock Market Cycles." *Review of Finance* 12 (2008): 253–92.

Kash, Don E. and Robert W. Rycroft. *US Energy Policy: Crisis and Complacency*, 1st edn. Norman: University of Oklahoma Press, 1984.

Katz, James Everett. *Congress and National Energy Policy*. New Brunswick, NJ: Transaction Books, 1984.

Katzenstein, Peter J. "Introduction: Domestic and International Forces and Strategies of Foreign Economic Policy." *International Organization* 31 (1977): 587–606.

Kaufman, Burton I. "Oil and Antitrust: The Oil Cartel Case and the Cold War." *The Business History Review* 51 (1977): 35–56.

Kaufmann, Robert K., Stephane Dees, Pavlos Karadeloglou, and Marcelo Sanchez. "Does OPEC Matter? An Econometric Analysis of Oil Prices." *The Energy Journal* 25 (2004): 67–90.

Keeler, John and Martin Shain, eds. *Chirac's Challenge: Liberalization, Europeanization, and Malaise in France*. Basingstoke: Macmillan Press, 1996.

Keidanren. *Keidanren 50-Nenshi [Fifty-Year History of Keidanren]*. Tokyo: Keidanren, 1999.

Kellam, R.O. "Regulation of Oil Imports." *Duke Law Journal* (1961): 175–215.

Kelly, William B., Jr. "The 'Expanded' Trade-Agreements Escape Clause 1955–61." *Journal of Political Economy* 70 (1962): 37–63.

Kemp, Murray C. and Ngo Van Long. "The Role of Natural Resources in Trade Models." *Handbook of International Economics* 1 (1984): 367–417.

Kenny, Ludwell. *We Fight for Oil*. New York: Alfred A. Knopf, 1928.

Keohane, Robert O. *After Hegemony: Cooperation and Discord in the World Political Economy*. Princeton University Press, 1984.

"State Power and Industry Influence: American Foreign Oil Policy in the 1940s." *International Organization* 36 (1982): 165–83.

"The Theory of Hegemonic Stability and Changes in International Economic Regimes, 1967–1977." In *Changes in the International System*, edited by Ole R. Holsti, Randolph Siverson, and Alexander George, 131–62. Boulder, CO: Westview, 1980.

Kikkawa, Takeo. "1934-nen no Nihon no Sekiyugyoho to Sutandado Bakyumu Oiru Kampani (1) [The 1934 Petroleum Industry Law and Standard Vacuum Oil Company (1)]." *Aoyama Keiei Ronshu* 23 (1989): 21–43.

"1934-nen no Nihon no Sekiyugyoho to Sutandado Bakyumu Oiru Kampani (2) [The 1934 Petroleum Industry Law and Standard Vacuum Oil Company (2)]." *Aoyama Keiei Ronshu* 24 no. 2 (1989): 63–100.

"1934-nen no Nihon no Sekiyugyoho to Sutandado Bakyumu Oiru Kampani (3) [The 1934 Petroleum Industry Law and Standard Vacuum Oil Company (3)]." *Aoyama Keiei Ronshu* 24 no. 3 (1989): 39–68.

"GHQ no Sekiyu Seisaku to Mejazu no Tainichi Senryaku [GHQ's Oil Policy and the Majors' Strategy Towards Japan]." *Aoyama Keiei Ronshu* 26 (1991): 265–84.

Nihon Sekiyu Sangyo no Kyosoryoku Kochiku [Building the Competitiveness of Japan's Oil Industry]. Tokyo: Nagoya Daigaku Shuppankai, 2012.

"Senzen Nhon no Karuteru [Pre-war Japanese Cartels]." *Aoyama Keiei Ronshu* 25 (1991): 73–98.

Kindleberger, Charles P. "The Rise of Free Trade in Western Europe, 1820–1875." *The Journal of Economic History* 35 (1975): 20–55.

Kitschelt, Herbert, Peter Lange, Gary Marks, and John D. Stephens. "Convergence and Divergence in Advanced Capitalist Democracies." In *Continuity and Change in Contemporary Capitalism*, edited by Herbert Kitschelt, Peter Lange, Gary Marks, and John D. Stephens, 427–60. Cambridge University Press, 1999.

Knight, Jack. "Explaining Neoliberalism." In *The Rise of Neoliberalism and Institutional Analysis*, edited by John L. Campbell and Ove K. Pedersen, 27–50. Princeton University Press, 2001.

"Models, Interpretations, and Theories: Constructing Explanations of Institutional Emergence and Change." In *Explaining Social Institutions*, edited by Jack Knight and Itai Sened, 95–120. Ann Arbor, MI: University of Michigan Press, 1995.

Kohl, Wilfrid L. *After the Second Oil Crisis: Energy Policies in Europe, America, and Japan*. Lexington, MA: Lexington Books, 1982.

Komiya, Ryutaro. "Introduction." In *Industrial Policy of Japan*, edited by Ryutaro Komiya, Masahiro Okuno, and Kotaro Suzumura, 11–13. Tokyo: Academic Press Japan, Inc., 1988.

Kosai, Yutaka. "The Reconstruction Period." In *Industrial Policy of Japan*, edited by Ryutaro Komiya, Masahiro Okuno, and Kotaro Suzumura, 25–43. Tokyo: Academic Press Japan, Inc., 1988.

Krasner, Stephen D. *Defending the National Interest: Raw Materials Investments and US Foreign Policy*. Princeton University Press, 1978.

"State Power and the Structure of International Trade." *World Politics* 28 (1976): 317–47.

Kuisel, Richard F. *Ernest Mercier: French Technocrat*. Berkeley, CA: University of California Press, 1967.

Labonte, Marc and Gail Makinen. "Energy Independence: Would it Free the United States from Oil Price Shocks?" Congressional Research Service RS20727 (2000).

Landeau, Jean-Francois G. *Strategies of US Independent Oil Companies Abroad*. Ann Arbor, MI: University of Michigan Press, 1979.

Laoyza, Norman V. and Raimundo Soto. "On the Measurement of Market-oriented Reforms." *World Bank Research Working Paper WPS 3371* (2004).

Lax, Howard L. *Political Risk in the International Oil and Gas Industry*. Boston: IHRDC, 1983.

Leamer, Edward E. *Sources of International Comparative Advantage: Theory and Evidence*. Cambridge, MA: MIT Press, 1984.

Le Dez, Morgan, "France's Oil Sources and Supply Networks (1861–1950)." In *Oil Producing Countries and Oil Companies*, edited by Alain Beltran, 15–32. Brussels: P.E. Lang, 2011.

Levy, Gilles Pierre. "The Relationships between Oil Companies and Consumer State Governments in Europe, 1973–82." *Journal of Energy and Natural Resources Law* 2 (1984): 9–29.

Levy, Jonah D. "Economic Policy and Policy-Making." In *Developments in French Politics 3*, edited by Alistair Cole, Patrick Le Gales, and Jonah D. Levy, 170–95. Basingstoke: Palgrave Macmillan, 2005.

The State after Statism: New State Activities in the Age of Liberalization. Cambridge, MA: Harvard University Press, 2006.

Tocqueville's Revenge: State, Society, and Economy in Contemporary France. Cambridge, MA: Harvard University Press, 1999.

Libecap, Gary D. and Steven N. Wiggins. "Contractual Responses to the Common Pool: Prorationing of Crude Oil Production." *The American Economic Review* 74 (1984): 87–98.

Lieber, Robert J. "Energy Policies of the Fifth Republic: Autonomy Versus Constraint." In *The Fifth Republic at Twenty*, edited by William G. Andrews and Stanley Hoffmann, 311–29. Albany, NY: State University of New York Press, 1981.

"Energy Policy and National Security: Invisible Hand or Guiding Hand?" In *Eagle Defiant: United States Foreign Policy in the 1980s*, edited by Kenneth A. Oye, Robert J. Lieber, and Donald Rothchild, 167–90. Boston, MA: Little, Brown and Company, 1983.

Lindberg, Leon N. *The Energy Syndrome: Comparing National Responses to the Energy Crisis*. Lexington, MA: Lexington Books, 1977.

Lohmann, Susanne and Sharyn O'Halloran. "Divided Government and US Trade Policy: Theory and Evidence." *International Organization* 48 (1994): 595–632.

Loriaux, Michael. *France after Hegemony: International Change and Financial Reform*. Ithaca: Cornell University Press, 1991.

Lucas, N.J.D. *Energy in France: Planning, Politics, and Policy*. London: Europa Publications for the David Davies Memorial Institute of International Studies, 1979.

"The Role of Institutional Relationships in French Energy Policy." *International Relations* 5 (1977): 87–121.

Luciani, Giacomo and Mario Salustri. "Vertical Integration as a Strategy for Oil Security." In *Strategic Positioning in the Oil Industry*, edited by Paul Stevens, 23–44. Abu Dhabi, UAE: Emirates Center for Strategic Studies and Research, 1998.

Luong, Pauline Jones and Erika Weinthal. "Rethinking the Resource Curse: Ownership Structure, Institutional Capacity, and Domestic Constraints." *Annual Review of Political Science* 9 (2006): 241–63.

Luszig, Michael. "The Limits of Rent Seeking: Why Protectionists Become Free Traders." *Review of International Political Economy* 5 (1998): 38–63.

Mabro, Robert. *OPEC and the World Oil Market: The Genesis of the 1986 Price Crisis*. Oxford University Press, 1986.

OPEC's Production Policies: How Do They Work?, Why Don't They Work? Oxford Institute for Energy Studies, 1989.

Matlary, Janne Haaland. *Energy Policy in the European Union*. New York: St. Martin's Press, 1997.

Maull, Hanns. "Oil and Influence: The Oil Weapon Examined." In *Economic Issues and National Security*, edited by Klaus Eugen Knorr and Frank N. Trager, 259–88. Kansas, KS: Regents Press of Kansas, 1977.

May, Peter J. "Policy Learning and Failure." *Journal of Public Policy* 12 (1992): 331–54.

McArthur, John H. and Bruce R. Scott. *Industrial Planning in France*. Cambridge, MA: Harvard College, 1969.

McGillivray, Fiona. *Privileging Industry: The Comparative Politics of Trade and Industrial Policy*. Princeton University Press, 2004.

McGowan, Francis. "The Single Energy Market and Energy Policy: Conflicting Agendas?" *Energy Policy* 17 (1989): 547–53.

McLean, John G. and Robert Haigh. *The Growth of Integrated Companies*. Norwood, MA: Plimpton Press, 1954.

Melby, Eric D.K. *Oil and the International System: The Case of France, 1918–1969*. New York: Arno Press, 1981.

Mendershausen, Horst. *Coping with the Oil Crisis: French and German Experiences*. Baltimore, MD: Johns Hopkins University Press, 1976.

Meritet, Sophie. "French Perspectives in the Emerging European Union Energy Policy." *Energy Policy* 35 (2007): 4767–71.

Meseguer, Covadonga. "What Role for Learning? The Diffusion of Privatization in OECD and Latin American Countries." *Journal of Public Policy* 24 (2004): 299–325.

Milner, Helen V. *Interests, Institutions, and Information: Domestic Politics and International Relations*. Princeton University Press, 1997.

Resisting Protectionism: Global Industries and the Politics of International Trade. Princeton University Press, 1988.

"Trading Places: Industries for Free Trade." *World Politics* 40 (1988): 350–76.

Milner, Helen V. and Keiko Kubota. "Why the Move to Free Trade? Democracy and Trade Policy in the Developing Countries." *International Organization* 59 (2005): 107–43.

Milner, Helen V. and David B. Yoffie. "Between Free Trade and Protectionism: Strategic Trade Policy and a Theory of Corporate Trade Demands." *International Organization* 42 (1989): 239–72.

Moe, Terry M. "Political Institutions: The Neglected Side of the Story." *Journal of Law, Economics, & Organization* 6 (1990): 213–53.

al-Moneef, Majed A. "Vertical Integration Strategies of the National Oil Companies." *Developing Economies* 36 (1998): 203–22.

Montague, Gilbert Holland. "The Rise and Supremacy of the Standard Oil Company." *The Quarterly Journal of Economics* 16 (1902): 265–92.

Moravcsik, Andrew. *The Choice for Europe: Social Purpose and State Power from Messina to Maastricht*. Ithaca: Cornell University Press, 1998.

Morse, Edward L. *Foreign Policy and Interdependence in Gaullist France*. Princeton University Press, 1973.

Morse, Ronald A. "Introduction: Japan's Energy Policies and Options." In *The Politics of Japan's Energy Strategy: Resources–Diplomacy–Security*, edited by Ronald A. Morse, 1–14. Berkeley, CA: University of California, 1981.

Morton, Rogers C.B. "The Nixon Administration Energy Policy." *Annals of the American Academy of Political and Social Science* 410 (1973): 65–74.

Moss, David A. *When All Else Fails: Government as the Ultimate Risk Manager*. Cambridge, MA: Harvard University Press, 2002.

Murat, Daniel, *L'intervention de l'État dans le secteur pétrolier en France [The Intervention of the State in the Petroleum Sector in France]*. Paris: Éditions Technip, 1969.

Nakatani, Iwao and Hiroko Ota. *Keizai Kaikaku no Bijon: Hiraiwa Ripoto Wo Koete [A Vision of Economic Reform: Beyond the Hiraiwa Report]*. Tokyo: Toyo Keizai Shinposha, 1994.

Nash, Gerald D. *United States Oil Policy 1890–1964*. University of Pittsburgh Press, 1968.

Nayberg, Roberto. "La politique française du pétrole à l'issue de la première guerre mondiale: perspectives et solutions." *Guerres mondiales et conflits contemporains* 4 (2006): 111–33.

Nelson, Garrison. *Committees in the US Congress, 1947–1992*. http://hdl.handle.net/1721.1/18166.

Neustadt, Richard E. *Presidential Power and the Modern Presidents: The Politics of Leadership from Roosevelt to Reagan*. New York: Maxwell Macmillan, 1990.

Nicolazzi, Massimo. "Petroleum Law and Petroleum Joint Ventures in Italy." *Journal of Energy and Natural Resources Law* 8 (1990): 30–50.

Noda, Tomio. "Nenryo Kokusaku to Sekiyu Shigen Kaihatsu: Kita Karafuto Sekiyu Kabushikigaisha to Teikoku Sekiyu Kabushikigaisha [National Fuels Policy and the Development of Oil Resources: The Kita Karufuto Oil Company and Imperial Oil Company]." *Journal of Political Economy* 70 (2004): 53–72.

Nordhauser, Norman E. "Origins of Federal Oil Regulation in the 1920s." *The Business History Review* 47 (1973): 53–71.

The Quest for Stability: Domestic Oil Regulation, 1917–1935. New York: Garland, 1979.

Noreng, Øystein. *Crude Power: Politics and the Oil Market*. London: I.B. Tauris, 2002.

North, Douglass C. *Institutions, Institutional Change, and Economic Performance*. Cambridge University Press, 1990.

Nowell, Gregory P. *Mercantile States and the World Oil Cartel, 1900–1939*. Ithaca: Cornell University Press, 1994.

O'Connor, Michael and James Bruining. "Regulation of Offshore Petroleum Activities in Ireland." *Journal of Energy and Natural Resources Law* 19 (2001): 132–42.

OECD. "IEA Work on Defining and Measuring Environmentally Harmful Subsidies in the Energy Sector." OECD Workshop on Environmentally Harmful Subsidies (2002).

Office of Technology Assessment. *US Oil Production: The Effect of Low Oil Prices – Special Report*, OTA-E-348. Washington, DC: US Government Printing Office, September 1987.

Olson, Mancur, Jr. "American Materials Policy and the Physiocratic Fallacy." *Orbis* 6 (1963): 670–88.

Oppenheimer, Bruce Ian. *Oil and the Congressional Process*. Lexington, MA: Lexington Books, 1974.

Oyama, Kosuke. "The Policymaking Process Behind Petroleum Industry Regulatory Reform." In *Is Japan Really Changing Its Ways? Regulatory Reform and the Japanese Economy*, edited by Lonny E. Carlile and Mark Tilton, 142–62. Washington, DC: Brookings Institution Press, 1998.

Padgett, Stephen. "The Single European Energy Market: The Politics of Realization." *Journal of Common Market Studies* 30 (1992): 53–76.

Pareto, Vilfredo. *Manual of Political Economy*. New York: A.M. Kelley, 1971 [1927].

Parra, Francisco R. *Oil Politics: A Modern History of Petroleum*. London and New York: I.B. Tauris, 2004.

"OPEC and the Price of Oil in 1993." *The Energy Journal* 15 (1994): 17–26.

Peltzman, Sam, Michael E. Levine, and Roger G. Noll. "The Economic Theory of Regulation after a Decade of Deregulation." *Brookings Papers on Economic Activity* (1989): 1–59.

Perry, Martin K. "Vertical Integration: Determinants and Effects." *Handbook of Industrial Organization* 1 (1989): 183–255.

Philippon, Andre. "The French Example: The 1928 Laws." In *A Comparative History of National Oil Companies*, edited by Alain Beltran, 21–54. Berlin, Germany: Fritz Lang, 2010.

Pierson, Paul. *Dismantling the Welfare State? Reagan, Thatcher, and the Politics of Retrenchment*. Cambridge University Press, 1994.

"Review: When Effect Becomes Cause: Policy Feedback and Political Change." *World Politics* 45 (1993): 595–628.

Pinto, Pablo M. and Santiago M. Pinto. "The Politics of Investment: Partisanship and the Sectoral Allocation of Foreign Direct Investment." *Economics & Politics* 20 (2008): 216–54.

Pollio, Gerald and Koichi Uchida. "Management Background, Corporate Governance and Industrial Restructuring: The Japanese Upstream Petroleum Industry." *Energy Policy* 27 (1999): 813–32.

Porter, Michael E. *The Competitive Advantage of Nations*. New York: Free Press, 1998 [1990].

Potters, Jan and Randolf Sloof. "Interest Groups: A Survey of Empirical Models that Try to Assess their Influence." *European Journal of Political Economy* 12 (1996): 403–42.

Priest, Tyler. *The Offshore Imperative: Shell Oil's Search for Petroleum in Postwar America*. College Station, TX: Texas A&M Press, 2007.

Prindle, David F. *Petroleum Politics and the Texas Railroad Commission*. Austin, TX: University of Texas Press, 1981.

Randall, Stephen J. "Harold Ickes and United States Foreign Petroleum Policy Planning, 1939–1945." *The Business History Review* 57 (1983): 367–87.

United States Foreign Oil Policy since World War I: For Profits and Security, 2nd edn. Montreal: McGill-Queen's University Press, 2005.

Ray, Edward John. "The Determinants of Tariff and Nontariff Trade Restrictions in the United States." *Journal of Political Economy* 89 (1981): 105–21.

Rodriguez, A.E. and Mark D. Williams. "Is the World Oil Market 'One Great Pool'? A Test." *Energy Studies Journal* 5 (1993): 121–30.

Rodrik, Dani. "Political Economy of Trade Policy." *Handbook of International Economics* 3 (1995): 1457–94.

"Understanding Economic Policy Reform." *Journal of Economic Literature* 34 (1996): 9–41.

Roggenkamp, M., C. Redgwell, I. Del Guayo, and A. Ronne. *Energy Law in Europe: National, EU and International Regulation*. New York: Oxford University Press, 2007.

Rogowski, Ronald. "Political Cleavages and Changing Exposure to Trade." *American Political Science Review* 81 (1987): 1121–37.

Rose, Andrew. "Do WTO Members have a More Liberal Trade Policy?" *NBER Working Paper* 9437 (2002).

Ross, Michael L. "What Do We Know about Natural Resources and Civil War?" *Journal of Peace Research* 41 (2004): 337–56.

Rouxel, Christian. *D'Azur à Total: Desmarais Frères, le premier grand pétrolier français*. Paris: Editions Drivers, 2007.

Ruggie, John Gerard. "International Regimes, Transactions, and Change: Embedded Liberalism in the Postwar Economic Order." *International Organization* 36 (1982): 379–415.

Saad, Farid Wadie. "France and Oil: A Contemporary Economic Study." PhD manuscript, Massachusetts Institute of Technology, 1969.

Samuels, Richard J. *The Business of the Japanese State: Energy Markets in Comparative and Historical Perspective*. Ithaca: Cornell University Press, 1987.

"Leadership and Political Change in Japan: The Case of the Second Rincho." *Journal of Japanese Studies* 29 (2003): 1–31.

Saul, Samir. "SN REPAL, CFP, and Oil Paid in Francs." In *A Comparative History of National Oil Companies*, edited by Alain Beltran, 93–123. Brussels: Peter Lang, 2010.

Schmidt, Vivien A. *From State to Market? The Transformation of French Business and Government*. Cambridge University Press, 1996.

"Loosening the Ties That Bind: The Impact of European Integration on French Government and Its Relationship with Business." *Journal of Common Market Studies* 34 (1996): 223–54.

Schneider, Steven A. *The Oil Price Revolution*. Baltimore, MD: Johns Hopkins University Press, 1983.

Seawright, Jason and John Gerring. "Case Selection Techniques in Case Study Research: A Menu of Qualitative and Quantitative Options." *Political Research Quarterly* 61 (2008): 294–308.

Selivanova, Yulia. *The WTO and Energy: WTO Rules and Agreements of Relevance to the Energy Sector*. Geneva: International Center for Trade and Sustainable Development, 2007.

Shaffer, Brian. "Firm-level Responses to Government Regulation: Theoretical and Research Approaches." *Journal of Management* 21 (1995): 495–514.

Shaffer, Ed. *The Oil Import Program of the United States: An Evaluation.* New York: Praeger, 1968.

Simmons, Beth A. and Zachary Elkins. "The Globalization of Liberalization: Policy Diffusion in the International Political Economy." *American Political Science Review* 98 (2004): 171–89.

Simmons, Beth A., Frank Dobbin, and Geoffrey Garrett. "Introduction: The International Diffusion of Liberalism." *International Organization* 60 (2006): 781–810.

Skowronek, Stephen. *Building a New American State: The Expansion of National Administration Capacities 1877–1920.* Cambridge University Press, 1982.

Smil, Vaclav. *Oil.* Oxford: Oneworld, 2008.

Smith, James L. "Inscrutable OPEC? Behavioral Tests of the Cartel Hypothesis." *The Energy Journal* 26 (2005): 51–83.

Soederberg, Susanne, Georg Menz, and Philip G. Cerny. *Internalizing Globalization: The Rise of Neoliberalism and the Decline of National Varieties of Capitalism.* Basingstoke: Palgrave Macmillan, 2005.

Sohn, Yul. *Japanese Industrial Governance: Protectionism and the Licensing State.* London: RoutledgeCurzon, 2005.

Solomon, Robert. *Money on the Move: The Revolution in International Finance Since 1980.* Princeton University Press, 1999.

Solow, Robert M. "The Economics of Resources or the Resources of Economics." *The American Economic Review* 64 (1974): 1–14.

Stagliano, Vito. *A Policy of Discontent: The Making of a National Energy Strategy.* Tulsa, OK: Pennwell, 2001.

Stern, Selma. "Turkey's Draft Petroleum Law." *Journal of Energy and Natural Resources Law* 23 (2005): 308–22.

Stevens, Paul. "Economists and the Oil Industry: Fact versus Analysis, the Case of Vertical Integration." In *Energy in a Competitive Market: Essays in Honour of Colin Robinson*, edited by Lester C. Hunt, 95–102. Cheltenham: Edward Elgar, 2003.

"A Survey of Structural Change in the International Oil Industry 1945–1984." In *The Changing Structure of the World Oil Industry*, edited by David Hawdon. London: Croom Helm, 1985.

"Understanding the Oil Industry: Economics as a Help or Hindrance." *The Energy Journal* 16 (1995): 125–37.

Stewart, Charles and Jonathan Woon. *Congressional Committee Assignments, 103rd to 112th Congresses, 1993–2011.*

Stigler, George J. "The Theory of Economic Regulation." *The Bell Journal of Economics and Management Science* 2 (1971): 3–21.

Stoff, Michael B. "The Anglo-American Oil Agreement and the Wartime Search for Foreign Oil Policy." *The Business History Review* 55 (1981): 59–74.

Oil, War and American Security: The Search for a National Policy on Foreign Oil 1941–1947. Hartford, CT: Yale University Press, 1980.

Streeck, Wolfgang and Kathleen Ann Thelen. *Beyond Continuity: Institutional Change in Advanced Political Economies*. Oxford University Press, 2005.

Suzuki, Yoshio. *Nihon no Kinyu Seisaku [Japanese Fiscal Policy]*. Tokyo: Iwanami Shinsho, 1993.

Tetreault, Mary Ann. *Revolution in the World Petroleum Market*. Westport, CT: Quorum Books, 1985.

Thelen, Kathleen Ann. *How Institutions Evolve: The Political Economy of Skills in Germany, Britain, the United States, and Japan*. Cambridge University Press, 2004.

Tietenberg, Thomas H. *Energy Planning and Policy: The Political Economy of Project Independence*. Lexington, MA: Lexington Books, 1976.

Toninelli, Pier Angelo. "Energy Supply and Economic Development in Italy: The Role of the State-Owned Companies." *Department of Economics Working Paper Series No. 146* (October 2008).

Turner, Louis. *Oil Companies in the International System*, 3rd edn. Winchester, MA: Allen & Unwin, 1983.

United Nations Conference on Trade and Development. *Trade Agreements, Petroleum and Energy Policies*. Geneva: United Nations, 2000.

VanDoren, Peter. *Politics, Markets, and Congressional Policy Choices*. Ann Arbor, MI: University of Michigan Press, 1991.

Varangu, Kristi. "IEA Work on Defining and Measuring Environmentally Harmful Subsidies in the Energy Sector." *OECD Workshop on Environmentally Harmful Subsidies*. Paris: OECD, 2002.

Venn, Fiona. *Oil Diplomacy in the Twentieth Century*. New York: St. Martin's Press, 1986.

Verleger, Philip K. *Adjusting to Volatile Energy Prices*. Washington, DC: Institute for International Economics, 1993.

Vernon, Raymond. *Two Hungry Giants : The United States and Japan in the Quest for Oil and Ores*. Cambridge, MA: Harvard University Press, 1983.

Vietor, Richard H.K. *Energy Policy in America since 1945: A Study of Business–Government Relations*. Cambridge University Press, 1984.

Vogel, Steven. *Freer Markets, More Rules: Regulatory Reform in Advanced Industrial Countries*. Ithaca: Cornell University Press, 1996.

Watkins, G.C. "Oil Scarcity: What Have the Past Three Decades Revealed?" *Energy Policy* 34 (2006): 508–14.

Weiner, Robert J. "Is the World Oil Market 'One Great Pool'?" *The Energy Journal* 12 (1991): 95–108.

Whitford, Andrew B. "Competing Explanations for Bureaucratic Preferences." *Journal of Theoretical Politics* 19 (2007): 219–47.

Wildavsky, Aaron. *The New Politics of the Budgetary Process*, 2nd edn. New York: Harper Collins, 1992.

Williamson, Harold Francis. *The American Petroleum Industry*. Evanston, IL: Northwestern University Press, 1959.

Williamson, Oliver E. "Transaction Cost Economics: The Governance of Contractual Relations." *Journal of Law and Economics* 22 (1979): 233–61.

Wilson, Ernest J. "World Politics and International Energy Markets." *International Organization* 41 (1987): 125–49.

Yamaoka, Akira. "Sekiyu Gyoho Kara Kyohan Gyosha Koso he [From the Petroleum Industry Law to the Joint Sales Company]." *Keizai Ronso* 170 (2002) 46–61.

Yergin, Daniel. "Ensuring Energy Security." *Foreign Affairs* 85 (2006): 69–82.

The Prize: The Epic Quest for Oil, Money, and Power. New York: Simon and Schuster, 1991.

Zeiler, Thomas W. "Kennedy, Oil Imports, and the Fair Trade Doctrine." *The Business History Review* 62 (1990): 286–310.

Zillman, Donald N. "Energy Trade and the National Security Exception to the GATT." *Journal of Energy and Natural Resources Law* 12 (1994): 117–27.

Zysman, John. *Governments, Markets, and Growth: Financial Systems and the Politics of Industrial Change*. Ithaca: Cornell University Press, 1983.

Political Strategies for Industrial Order: State, Market, and Industry in France. Berkeley, CA: University of California Press, 1977.

Index

action spécifique, *see* golden share
advanced industrialized states, oil market governance in, 30–2
Africa, oil production in, 30
AGIP, *see* Azienda Generale Italiano Petroli
Algeria, French oil market and, 85
American Petroleum Institute (API), 173, 175–7, 189–90
ANWR, *see* Arctic National Wildlife Reserve
API, *see* American Petroleum Institute
Arctic National Wildlife Reserve (ANWR), 185–6
AREVA, 70
Article 8, 82
Asia, *see also specific nations*
  oil production in, 29–30
Azienda Generale Italiano Petroli (AGIP), 31

Bentsen, Lloyd, 178–9, 182
Bérégovoy, Pierre, 99
Boulter, Eldon, 181–2
Bowles, Jim L., 190
Breaux, John, 186
BRP, *see* Bureau de recherches de pétrole
Bryant, John, 181
Bureau de recherches de pétrole (BRP), 31, 76
Burtis, Theodore, 149, 180–1
Bush, George W., presidency, 170, 185–7

Callon Petroleum, 181
CFP, *see* Compagnie Française des Pétroles
CFR, *see* Compagnie Française de Raffinage
Chafee, John, 182–3
Chalandon, Albert, 94
Chevron, 175–6, 188, 190
China, market, 18–19, 207–8
Chirac, Jacques, 96
Churchill, Winston, 46
Compagnie Française de Raffinage (CFR), 31, 74
  expansion of, 85
Compagnie Française des Pétroles (CFP), 31, 73–4
  expansion of, 84–5
  nationalization as influence on, 85–6, *see also* CFP/Total
Connally Act (US), 158
Conoco, 155
Coolidge, Calvin, 46

Daikyo Oil, 133
Davies, Ralph, 160
Deepwater Royalty Relief Act (US), 189
Desmarais Frères, 71
DiBona, Charles, 166, 174, 179–80
Directorate of Hydrocarbons, in France, 98–101

EC, *see* European Commission
ECJ, *see* European Court of Justice
Economic Recovery Act (US), 173
EDF, *see* Électricité de France
Edwards, James, 183
80/20 rule, 101–2, *see also* France, import controls in
Électricité de France (EDF), 70
Elf Aquitaine, 63, 66
  diversification of, 83
  oil market governance and, 88–95
  privatization of, 68, 92–5, 102

Emergency Petroleum Allocation Act (EPAA) (US), 165–6
Energy Policy Act (US), 193–4
Energy Security Act (US), 178–9, 182
energy sources, *see* nuclear energy
Ente Nazionale Idrocarburi (ENI), 31
Entreprise de recherches et d'activités pétrolières (ERAP), 77
EPAA, *see* Emergency Petroleum Allocation Act
ERAP, *see* Entreprise de recherches et d'activités pétrolières
Europe, oil market governance in, 61–2, *see also* France, oil market and
  institutional changes in, 201
  OECD strategies for, 61–2
European Commission (EC)
  French oil market influenced by, 78–81, 104–10
  SEM and, 107
European Court of Justice (ECJ), 109–10
Exxon, 154–5, 175–6, 188

Fabius, Laurent, 99
Federal Oil Conservation Board (FOCB), 157–8
financial liberalization, 45
firm characteristics
  diversification of, 66–7, 198–9
  in France, 198–9
  industrial compacts and, 49–50, 199–204
  in Japan, 115, 199
  liberalization and, 14, 53, 198–9
  origins of, 16, 55–8, 155, *see also* industrial compacts; resource endowments
  state ownership of, 35
  in United States, 166–71
firms, in oil market
  competition between, in US, 3
  in France, strategies for, 83–102
  government role in, 15–17, 35, 51
  governments influenced by, 25–6, 54–5
  interests and strategies of, 47, 50–5
  internationalized, 14, 25–6, 203
  in Japan, 114, 115
  majors, 26, 47, *see also* majors, oil
  in US, 3, 155, 166–71, 174–85, 188–93
  vertical integration in, 25–6, 50–1, 78, 202
FOCB, *see* Federal Oil Conservation Board
Follis, R. G., 163
formula-based prices, for crude oil, 52
France, oil market and
  Algerian oil production, 85
  Article 8, 82
  BRP, 31, 76
  CFP/Total, 31, 73–4, 84–6, *see also* Total
  CFR, 31, 74, 85, *see also* CFP/Total
  colonialism, 62
  Desmarais Frères, 71
  Directorate of Hydrocarbons, 98–101
  diversification, 86–7
  early imports, 70–1
  EC influence, 78–81, 104–10
  ECJ rulings, 109–10
  EDF, 70
  80/20 rule, 101–2
  energy laws, 68
  ERAP, 77
  firm strategies, 83–102
  golden share, use of, 97, 109–10, 211, 215–17
  government involvement, 16, 77, 95–102
  historical development, 70–8
  ideational change, 110–12
  IEA, 110
  import controls, 73, 81, 91, 92, 99–100
  industrial compacts, 58, 72–4, 75–6
  institutional changes, 202
  Law of 1928, 103
  laws, 82
  liberalization, 68, 69–70, 90–1
  market competition within, 71, 83–8
  market volatility, 80, 88
  Ministry of Finance, 98–9, 110
  90/10 rule, 101, *see also* France, oil market and import controls

nuclear energy, 63, 70
oil market governance, 30–2, 70–8, 81–3, 88–95
ONCL, 72–3
price formation, 74–5, 81, 90–1, 99
privatization, 81–2, 92–5
refinery performance, 98
resource nationalism, 207–9
Royal Dutch Shell, 71–2
SEP initiative, 72
SNPA, 76, 84
Standard Oil, 71–2
Treaty of Rome, 69
Treaty of San Remo, 72
Union générale des petroles (UGP), 31, 76
US oil market, 150
vertical integration of firms, 78

Gaz de France, 70
General Agreement on Trade and Tariffs (GATT)
oil market governance under, 6, 41–2
Punta del Este Declaration, 41–2
security interests under Article XXI, 42
golden share, 97, 109–10, 211, 215–17
governments, oil markets and
demand and supply influenced by, 15–17, 51
distributive effects, 18
firms as influence on, 25–6
firms influenced by, 15–17, 51
France, 16, 77, 95–102
international firms as influence on, 26
international security, 18
Japan, 16–17
market share influenced by, 51
political partisanship, 44–5
Standard Oil, regulations for, 157–8
state ownership, 35
US, 17, 152–4
Gulf Oil, 175

Hamilton, Charles, 159
Hatch, Orrin, 192
Hefner, Raymond, 177
Hervé, Edmond, 90
Hiraiwa Commission, 136–7
Hotelling, Harry, 59–60
Hubbert, M. King, 60

ideational change
and French oil market, 110–12
and Japanese oil market, 146–8
and oil market governance, 58–62, 200–1
and US oil market, 196–7
Idemitsu Kosan, 123
IEA, *see* International Energy Agency
Imperial Oil, 120, 124, 143
import controls
in French oil market, 81, 91
in Japanese oil market, 140
in US oil market, 168–9, 170–1
import quotas, 73
imports, in oil markets, *see also* Middle East, oil production in
French oil market and, 70–1
from US, 25
Independent Petroleum Association of America (IPAA), 154, 161–2, 164, 177–8, 202
India, as new oil market, 207–8
industrial compacts
early negotiation history for, 200
in France, 58, 72–4
in Japan, 58, 114, 206–7
liberalization of oil markets and, 206–7
oil market governance and, 49–50, 199–204
in US, 167–8
innovation
in manufacturing, 7
in oil markets, 2
institutional changes, in oil markets
in Europe, 201
in France, 202
in US, 202–3
institutions, 64–5, 69–70
International Energy Agency (IEA)
French oil market and, 110
functions of, 7, 42
oil market governance and, 7
Shared Goals, 32–3
transaction costs and, 7
US oil market and, 150, 184

international organizations, 7, 41–3, *see also* General Agreement on Trade and Tariffs (GATT); International Energy Agency (IEA); Organisation of Economic Co-operation and Development (OECD); Organization of Petroleum Exporting Countries (OPEC); World Trade Organization (WTO)
IPAA, *see* Independent Petroleum Association of America
Iraqi National Oil Corporation, 28
Italy
 AGIP, 31
 ENI, 31
 oil market governance in, 31

Japan, oil market in
 cartels as part of, 118–19
 consumer demands in, 138–9
 Cosmo Oil, 133
 Daikyo Oil, 133
 diversification of, 128–9, 199
 exploration efforts in, 119–20
 firms in, 114, 115
 government influence on, 16–17, 117–18
 Hiraiwa Commission, 136–7
 historical development of, 115–21
 horizontal fragmentation in, 126, 134
 ideational change and, 146–8
 Idemitsu Kosan, 123
 Imperial Oil, 120, 124, 143
 importing controls, 140
 industrial compacts in, 58, 114, 206–7
 INPEX, 143
 JNOC, 130–1, 143, 148
 JOGMEC, 143–4
 JPDC, 125
 Kyodo Petroleum, 124, 133
 lack of diversification in, 130, 134–5
 Law Concerning Foreign Investment, 121
 Liberal Democratic Party in, 140–1
 Maekawa Commission, 136
 Maruzen, 133
 mergers within, 124, 141–2
 Nihon Seikiyu, 116–17, 128–9
 under Nine Power Treaty of 1922, 119
 nuclear energy and, 63
 oil laws in, 119–20
 oil market governance in, 30–2, 115–27
 price volatility in, 129, 131–2, 142
 privatization of, 114
 promotion of mergers in, 124
 refinery and production in, 114, 120–1, 130
 resource nationalism and, 207–9
 Rising Sun, 116
 Royal Dutch Shell in, 116–17
 SCAP influence on, 120–1
 Socony Vacuum, 116
 Standard Oil and, 116–17
 Toyo Oil, 135–6
 US oil market compared to, 150
 after World War II, 121–8
 during World War II, 120–1
Japan National Oil Company (JNOC), 130–1, 143, 148
Japan Oil, Gas and Metals National Company (JOGMEC), 143–4
Japan Petroleum Development Corporation (JPDC), 125
JNOC, *see* Japan National Oil Company
Johnston, J. Bennett, 182
JPDC, *see* Japan Petroleum Development Corporation

Keller, George M., 175
Kunihiko, Matsuo, 143–4
Kuwait Petroleum Corporation, 28

Latin America, oil production in, 29
Law Concerning Foreign Investment (Japan), 121, 123–4
Law of 1928 (France), 103
Liberal Democratic Party (Japan), 140–1
liberalization, of oil markets
 definition of, 10
 economic crises as factor in, 9
 firm competitiveness and, 204–5
 in France, 69–70, 90–1
 industrial compacts and, 206–7
 in Japan, 122

market governance and, 9–10, 35–6, 40, 204–7, *see also* oil market governance
measurement of, 35–6
sequencing and, 205–6
transaction costs and, 42–3
licensing, *see* import controls
Loïk, Le Floch-Prigent, 86–7, 94

MacFarlane, Charles, 189–90
Madelin, Alain, 101
Maekawa Commission, 136
majors, oil, 25–6, 47, 52, 116–17, 122, *see also* vertical integration, in international firms
Martin, William, 183–4
Mer, Francis, 110
Middle East, oil production in, *see also* Organization of Petroleum Exporting Countries; Saudi Arabia, market share for
competitive market instability as result of, 28
growth and development of, 24–5
Standard Oil and, 163
US oil market influenced by, 152, 161–6
Mineral Leasing Act (US), 159
Mitchell, George J., 183
Mitterand, François, 96–7
Mobil, 176
Morozumi, Yoshihiro, 141
Murkowski, Frank, 186

Naito, Masahisa, 139–40
National Energy Plan (NEP), 186–7
National Industry Recovery Act (US), 158
National Petroleum Security Act (US), 181–2
nationalization, of oil production, 15
NEP, *see* National Energy Plan
New International Economic Order (NIEO) movement, oil market governance during, 6
new markets, for oil, 207–8
Nickles, Don, 182
NIEO movement, *see* New International Economic Order movement
Nigerian National Petroleum Corporation, 28
Nihon Seikiyu, 116–17
Nine Power Treaty of 1922 (Japan), 119
90/10 rule, 101, *see also* France, oil market and import controls
non-renewable resource, oil as, 12
nuclear energy
in France, 63, 70
in Japan, 63

OCS, *see* Outer Continental Shelf
OECD, *see* Organisation of Economic Co-operation and Development
Office Nationale des Combustibles Liquides (ONCL), 72–3
oil market governance, 5–15
in advanced industrialized states, 30–2
in China, 207–8
definition of, 32–3
demand and supply elements in, 49–50
domestic interests and, 44–9, *see also* firms, in oil market; governments, oil markets and; industrial compacts
Elf Aquitaine and, 88–95
in Europe, 61–2, *see also* European Commission; European Court of Justice
firms as influence on, 9–15, 26, 47
in France, 30–2, 70–8, 81–3
under GATT, 6, 41–2
ideational change and, 61–2, 200–1
IEA and, 7
in India, 207–8
industrial compacts and, 49–50, 199–204
by international firms, 26, 47, 198–9
in Italy, 31
in Japan, 30–2, 115–21
liberalization as, 7, 9–10, 35–6, 40, 204–7
measurement of, 33–7, 65–7
through nationalization, 15
during NIEO movement, 6
oil price declines and, 171–4

oil market governance (*cont.*)
political partisanship as influence on, 44–5
Shared Goals policy of, 32–3
statecraft and, 11–15
technology for, 2
Total and, 88–95
trade barriers and, 35, 41–2
in US, 30–2, 166–71, 194–6
world economy influenced by, 17–19
under WTO, 6
oil markets, production for, *see also* France, oil market and; governments, oil markets and; Japan, oil market in; Middle East, oil production in; United States (US), oil market in; *specific continents*; *specific nations*
barriers to, 35, 41–2
in China, 207–8
competitive market instability in, 28
distribution of, 26
diversification of production in, 25, 29–30
fragmentation of, 52
geological influence on, 11
global markets in, emergence of, 13, 24–6, 27–30
in India, 207–8
innovation in, 2
instability of prices in, 28–9, *see also* prices, for crude oil
international market for, *see* global markets in, emergence of
nationalization of, 15
OCS and, 153
in OECD, 56
OPEC and, formation of, 27–8
price volatility and, 52–3
production, in US, 152, 207
projections for, in US, 207
stability of prices in, 27
transportation applications and, historical shift towards, 24–5
vertically integrated international firms and, control of, 25–6
Omnibus Trade and Competitiveness Act (US), 173
ONCL, *see* Office Nationale des Combustibles Liquides
OPEC, *see* Organization of Petroleum Exporting Countries
Organisation of Economic Co-operation and Development (OECD)
in Europe, oil market governance strategies, 61–2
oil market governance and, 14
oil production share by, 56
Organization of Petroleum Exporting Countries (OPEC)
formation of, 27–8, 42
market share for, 28
share of global production by, decline of, 29
Outer Continental Shelf (OCS), 153, 173, 178, 190

Petróleos de Venezuela, S.A. (PDVSA), 28
Petroleum Reserves Corporation, 160–1
power, economic, 7, 18–19, 43–4
prices, for crude oil
diversification of supplies as factor in, 29
fall of, 171–4
fall of, oil market governance and, 171–4
formula-based, 52
in France, fixing of, 74–5, 81, 90–1
instability of, 28–9, 30, 52–3
in Japanese oil market, 129, 131–2, 142
liberalization of, in France, 90–1
oil market structure as influence on, 52–3
stability of, 27
in US oil market, abolition of controls, 165–6, 171–4
privatization
of Elf Aquitaine, 68, 92–5, 102
of French oil market, 81–2
in Japanese oil market, 114
of Total, 68, 92–5, 102
protectionism
in oil market governance, 5–6
in US oil market, 2, 151, 161–6

Punta del Este Declaration (GATT), 41–2

Raymond, Lee, 188
Rayner, Charles, 159–60
Reagan presidency, 171–85
refinery and distribution performance, 26
  in France, 98
  in Japan, 114, 120–1
resource endowments, 13–14, 55–7, 150–1, 166–9
  resource nationalism, 1–2, 44, 207–9
  in US oil market, 4, 150–1, 166–7
Richardson, Bill, 181
Rising Sun, 116
risk, governments as managers of, 10–11
Royal Dutch Shell
  in France, 71–2
  in Japan, 116–17
du Rusquec, Jean, 91, 95

Saudi Arabia, market share for, 28
SCAP, *see* Supreme Commander of Allied Powers
Scheer, François, 106
Securing America's Future Energy Act (US), 186
SEM, *see* Single European Market
SEP, *see* Société pour l'exploitation des pétroles
sequencing, 3, 205–6
Shared Goals policy, of IEA, 32–3
Single European Market (SEM), 107
Société nationale de pétroles (SNPA), 76
  expansion of, 84
Société pour l'exploitation des pétroles (SEP), 72
Socony Vacuum, 116
Sonatrach, 28, 85
Soviet Union, oil production rates in, 30
Standard Oil, 156–8, *see also* Exxon; majors, oil
  dissolution of, 157
  in France, 71–2
  in Japan, 116–17
  Middle East oil production and, 163
  as monopoly, 156
  US government regulation of, 157–8
Supreme Commander of Allied Powers (SCAP), 120–1

Tauzin, Billy, 186
Tax Reform Act (US), 184
technology, for oil market governance, 7, *see also* innovation, in oil markets
Texaco, 176
Texas Railroad Commission, 158
Thomas, Craig, 191
Total, 63, 66, *see also* CFP/Total; CFR
  diversification of, 83
  expansion of, 85
  oil market governance and, 88–95
  privatization of, 68, 92–5, 102
Toyo Oil, 135–6
Trade Agreements Extension Act (US), 162, 164
Trade Expansion Act (US), 173
transaction costs, 7, 42–3
transport costs, for oil markets, 2
Treaty of Rome, French oil market under, 69
Treaty of San Remo, France under, 72

Union générale des pétroles (UGP), 31, 76
United States (US), oil market in, 24, *see also* Standard Oil
  ANWR and, 185–6
  bureaucratic capacity in, 46
  during Bush presidency, 170, 185–7
  Callon Petroleum, 181
  case studies for, 171–4, 185–7
  Chevron, 175–6, 188
  under Connally Act, 158
  Conoco, 155
  under Deepwater Royalty Relief Act, 189
  depletion fears, 159
  diversification of, 152
  domestic production, 152, 156–8
  early production in, 4, 24
  under Economic Recovery Act, 173
  under Energy Policy Act, 193–4

United States (US), oil market in (*cont.*)
under Energy Security Act, 178–9, 182
under EPAA, 165–6
Exxon, 154–5, 175–6, *see also* majors, oil; Standard Oil
firm characteristics, 166–71
firm competitiveness, 3
firm heterogeneity in, 155
firm strategies, 174–85, 188–93
FOCB and, 157–8
French oil market compared to, 150
government influence on, 17, 152–4
history of, to 1980, 154–66
ideational philosophy within, 196–7
IEA and, 150, 184
import controls in, 168–9, 170–1
industrial compacts in, 167–8
institutional changes as influence on, 202–3
IPAA and, 154, 161–2, 164, 177–8, 202
Japanese oil market compared to, 150, 170–1
kerosene demand in, 24
liberalization of, 164–5
Middle East oil production as influence on, 152, 161–6
under Mineral Leasing Act, 159
under National Industry Recovery Act, 158
under National Petroleum Security Act, 181–2
under NEP, 186–7
as net importer, 25
OCS and, 153, 173
oil market governance in, 30–2, 166–71, 194–6
oil-producing states, 190–2
under Omnibus Trade and Competitiveness Act, 173
Petroleum Reserves Corporation in, 160–1
price controls for, abolition of, 165–6, 171–4
production projections for, 207
promotion of production, 158–61
protectionism in, 151, 161–6
public opinion, as influence on, 154
rates of production, 152, 207
during Reagan presidency, 171–85
regional associations, 177, 179
resource endowments in, 4, 150–1, 166–7, 207–9
sanctions in, 2
under Securing America's Future Energy Act, 186
under Tax Reform Act, 184
under Trade Agreements Extension Act, 162, 164
under Trade Expansion Act, 173
vertically integrated firms in, 202
Windfalls Profit Tax and, 173, 185
during World War II, 160–1
Unsell, Lloyd, 177, 178

vertical integration, in international firms, 25–6, 50
in France, 78
function of, 50–1
in US, 202
Vursell, Charles, 162

wars, oil market governance and, 46, *see also* World War I; World War II
Windfalls Profit Tax, 173, 185
World Trade Organization (WTO), oil market governance under, 6
World War I, 46
World War II
Japanese oil market during, 121–8
oil market after, in Japanese, 121–8
US oil market during, 160–1
WTO, *see* World Trade Organization